Lecture Notes in Mathematics

Edited by A. Dold and B. Eckmann

613

E. Behrends R. Danckwerts
R. Evans S. Göbel P. Greim
K. Meyfarth W. Müller

L^p-Structure in Real Banach Spaces

Springer-Verlag
Berlin Heidelberg New York 1977

Authors
Ehrhard Behrends
Rainer Danckwerts
Richard Evans
Silke Göbel
Peter Greim
Konrad Meyfarth
Winfried Müller

I. Mathematisches Institut
der Freien Universität Berlin
Hüttenweg 9
1000 Berlin 33/BRD

AMS Subject Classifications (1970): XX-46

ISBN 3-540-08441-X Springer-Verlag Berlin Heidelberg New York
ISBN 0-387-08441-X Springer-Verlag New York Heidelberg Berlin

Printed in Germany

Printing and binding: Beltz Offsetdruck, Hemsbach/Bergstr.
2141/3140-543210

C O N T E N T S

Introduction

Chapter 0: Preliminaries

Chapter 1: L^p-Projections

Chapter 2: The Cunningham p-Algebra

Chapter 3: The Integral Module Representation

Chapter 4: The Classical L^p-Spaces

Chapter 5: Integral Modules and Duality

Chapter 6: Spectral Theory for L^p-Operators

Chapter 7: The L^p-Structure of the Bochner Spaces and Related Results

Appendix 1: The Commutativity of L^p-Projections

Appendix 2: L^∞-Summands in CK-Spaces

Appendix 3: A Measure-Theoretical Approach to Integral Modules

Notation Index

Subject Index

References

Introduction

In 1972, Alfsen and Effros ([AE]) published a paper in which they discussed certain problems concerning the isometrical structure of Banach spaces. Using this paper as a starting point, a research group at the Freie Universität Berlin has been working on structural problems of this type since the spring of 1973. During this period the results achieved by the group have been published as papers, theses, and preprints. The purpose of these notes is to give the reader a more or less complete account of those results which can be grouped under the heading "L^p-structure".

Let X be a real Banach space and $1 \leq p \leq \infty$. Two closed subspaces J, $J^\perp$ of X are called complementary L^p-summands if X is the algebraic sum of J and $J^\perp$ and for every $x \in J$, $x^\perp \in J^\perp$

$$\|x + x^\perp\|^p = \|x\|^p + \|x^\perp\|^p \quad (\text{if } 1 \leq p < \infty)$$
$$\|x + x^\perp\| = \max\{\|x\|, \|x^\perp\|\} \quad (\text{if } p = \infty),$$

i. e. when the elements in J and $J^\perp$ behave like disjoint elements in an L^p-space. The projection from X onto J corresponding to this decomposition of X is called an L^p-projection and the set of all projections obtained in this way $\mathbb{P}_p$.

L^1- and L^∞-summands and the corresponding projections were first studied by Cunningham ([C1] and [C2]). Alfsen and Effros carried on the investigation in the above-mentioned paper, in which probably the most important results are the characterization of M-ideals by means of an intersection property (an M-ideal is a closed subspace whose polar is an L^1-summand in the dual space) and the introduction of the structure topology, with whose help one can prove a very genere-lized form of the Dauns-Hofmann theorem. They also applied the concepts to the most important concrete Banach spaces.

The general case, i. e. p not necessarily =1 or = ∞, has hardly been investigated at all, apparently. The main reason is that even very simple questions (e. g. Is the intersection of two L^p-summands also an L^p-summand) can only be answered if it is known that every pair of L^p-projection commute . In the case of $p = 1$ or $p = \infty$ this is easily seen and can also be proved directly for Banach spaces where the Clarkson inequality (cf. [L] , p. 169) is valid for the relevant p. Behrends has shown in [B2] that the answer to this question is affirmative for all $p \neq 2$. Since any orthogonal projection in a Hilbert space is an L^2-projection this result does not hold for p=2. On the other hand some important results hold for any complete Boolean algebra of L^p-projections, not necessarily containing them all. In chapters 3 - 5 we therefore formulate these results in the general context, which in particular means that we can apply them to the case $p = 2$ by considering maximal families of commuting projections.

Some authors ([CS],[E1]) have studied a natural generalization of the concept of L^p-summand. Let F be a mapping from $\mathbb{R}_+ \times \mathbb{R}_+$ into $\mathbb{R}_+$. We call two subspaces J, $J^\perp$ of a Banach space X, F-summands, if X is the algebraic direct sum of J and $J^\perp$ and further $F(\|x\|, \|x^\perp\|) =$

$\|x + x^{\perp}\|$ for all x in J, $x^{\perp}$ in $J^{\perp}$. ([CS] considers the special case $F(s,t) = f^{-1}(f(s) + f(t))$ for a continuous strictly monotone function $f: \mathbb{R}_+ \to \mathbb{R}_+$.) It can be shown (see note at the end of chapter 1) that if there are two nontrivial F-summands, one contained in the other, then $F = F_p$ for some p in $[1,\infty]$ whereby $F_p(s,t) = (s^p+t^p)^{1/p}$ for $p<\infty$ and $F_\infty(s,t) = \max\{s,t\}$. In this case F-summands are of course L^p-summands so that a restriction of our consideration to the latter does not involve any real loss of generality.

The main problem for choosing the material for these notes was that, while in the case of $p \neq 1, 2, \infty$, the definitions, propositions, proofs etc. are formally identical (differing only in the value of p) for all p, in the case $p = 1$ and $p = \infty$ the propositions are often only valid in a modified form or cannot be proved by the same method as in the general case. In the interest of uniformity we have therefore only mentioned those results which can be proved in more or less the same way as in the general case. (some results which we have left out for this reason can be found in [DGM]). The difference in the behaviour of the case $p = 1$ and $p = \infty$ as opposed to the other values of p is basically due to the fact that L^p-projections in dual spaces are necessarily w^*-continuous for $p > 1$ but not for $p = 1$ (which in particular means that M-ideals are not necessarily L^∞-summands). This result was proved independently by [F] and [E1] (in [E1] in a more general form for dual F-projections) although the method of the proof is the same as in [CER], who only consider the case $p = \infty$.

These notes fall into two main parts - chapters 1-3 in which the theory is developed and chapters 4-7 which deal with some applications. The contents of the individual chapters are as follows:

Chapter 1: The concept of an L^p-summand is explained with the help of some concrete examples. Although the proof of a lemma concerning the effect of transposition to L^p-summands and -projections is given in full, the main theorem concerning the commutativity of L^p-projections is only stated. A sketched proof can be found in appendix 1. It is shown that, for $p \neq 2$, $\mathbb{P}_p$ is a Boolean algebra and, for $p < \infty$, a complete one in which increasing nets converge pointwise to their suprema.

Chapter 2: The Cunningham p-algebra $C_p(X)$ (closure of the linear hull of $\mathbb{P}_p$ in $[X]$) and the Stonean space Ω of $\mathbb{P}_p$ (whose clopen subsets represent $\mathbb{P}_p$) are defined and examined. In particular it is shown that the Cunningham p-algebra is isomorphic in all structures to the space of continuous functions on Ω. The effect of taking products and quotients is also investigated.

Chapter 3: In the first part of the chapter we show how a Banach space X can be embedded in a field of Banach spaces over Ω in such a way that the L^p-projections in X have the effect of characteristic projections. This embedding (p-integral module representation) turns out to be the most important aid in the investigation of L^p-structure.

The second part contains some important consequences which are needed in the following chapters.

Chapter 4: With the help of the techniques of chapter 3 we show that abstract L^p-spaces can be characterized by the maximality of their L^p-structure - a Banach space X is isometric to an L^p-space if and only if $(C_p(X))_{COMM} = C_p(X)$. The most important result used in the proof is a lemma concerning the existence of projections in $(C_p(X))_{COMM}$ which generalizes a result of Cohen-Sullivan ([CS]) for smooth reflexive spaces.

We also give an explicit description of the L^p-summands in an L^p-space. It turns out that every L^p-summand is more or less the annihilator of a measurable set, a result already obtained in [Su 2].

Chapter 5: We study the relationship between the p-integral representation of a Banach space and the p'-integral representation of the dual $(1/p + 1/p' = 1)$, in particular the connection between the reflexivity of the space itself and that of the component spaces in the representation.

Chapter 6: In an analogous manner to the theory of self-adjoint operators in Hilbert space we represent the operators in the Cunningham p-algebra as Stieltjes integrals over spectral families of projections. It is shown that there is a 1-1 correspondence between the operators in $C_p(X)$ and normalized spectral families. We then give some important results in the general theory which follow from this.

Chapter 7: In this chapter we apply the representation of chapter 3 to some simple vector-valued L^p-spaces and draw some parallels to the general case.

In chapter 0 we have collected those results from other branches of mathematics which the reader will need to understand the following chapters. The appendices contain a sketched proof of theorem 1.3 (for a complete proof see [B2]), some remarks concerning the structure of the L^∞-summands in CK-spaces, and a discussion of a measure theoretic approach to integral modules.

It is clear that when a group have been working together for several years it is impossible to say which member is responsible for each result. Without forgetting this we can say however that the contributions of the individual members of the group are roughly as follows:- Chapter 1: Behrends ; Chapter 2: Danckwerts, S. Göbel, Meyfarth ; Chapter 3: Evans (section F together with Greim) ; Chapter 4: Evans ; Chapter 5: Greim ; Chapter 6 : Müller ; Chapter 7: Evans, Greim.

In conclusion we would like to thank the FNK (Kommission für Forschung und wissenschaftlichen Nachwuchs) of the Freie Universität Berlin for assisting us financially in the years 1974-75.

Chapter 0: Preliminaries

Topology

We assume that the reader is familiar with the elementary concepts of topology.

A nowhere dense set is a set A in a topological space such that the interior of the closure of A is empty. A set which is the union of countably many nowhere dense sets is said to be of first category. In a compact space ("compact" always includes the Hausdorff property) the empty set is the only open set of first category.

A topological space is said to be extremally disconnected when the closure of each open set is also open. If the space is also Hausdorff this implies that the connected components consist only of single points or in other words that the space is totally disconnected.

Consider the collection of sets of the form $A \Delta B$ where A is a clopen set and B is a set of first category. This collection is clearly closed under finite unions and intersections. If the space is extremally disconnected it is also closed under countable unions since the union of countably many sets of first category is also of first category and the union of countably many clopen sets is open and thus, since its closure is clopen differs from a clopen set by a set of first category. Since the complement of such a set also has this form it follows that in an extremally disconnected space the sets of this form form a σ-algebra. This σ-algebra contains the open sets since the closure of an open set is clopen and the boundary is of first category. In a compact space the equality $A \Delta B = C \Delta D$ (A, C clopen, B,D of first category) implies that $A = C$ and $B = D$ since $\emptyset$ is the only clopen set of first category.

A subset A of a topological space is called regularly closed if the closure of the interior of A is A. Since the interior of A is open it follows that in an extremally disconnected space the regularly closed sets are the clopen sets. In an extremally disconnected compact space the closure of an open set is homeomorphic to its Stone-Čech-compactification ([Sch], II.7.1).

Borel measures

In a topological space the Borel sets are the members of the σ-algebra generated by the open sets. It follows from the first part of this chapter that the Borel sets in a compact extremally disconnected space can be uniquely represented as the difference of a clopen set and a set of first category. A Borel measure is a σ-additive set function defined on the Borel sets. The support of a Borel measure m, denoted by supp m, is the set of all points such that every neighbourhood of the point contains a set with non-zero measure. The support is always closed. A Borel measure is said to be regular (from inside) if the measure of each set is the limit of the measures of the compact sets contained in it. The support of a regular Borel measure is regularly closed. The Riesz representation theorem states that the space of all finite regular Borel measures on a compact space is the dual of the space of continuous functions on this space (with the sup-norm) under the duality $<\mu,f> := \int f d\mu$. A regular content is a set function defined on the compact subsets with the following properties:

(i) $0 \leq m(D) < \infty$

(ii) $C \subset D$ implies $m(C) \leq m(D)$

(iii) $m(C \cup D) \leq m(C) + m(D)$

(iv) $m(C \cup D) = m(C) + m(D)$ for disjoint C, D

(v) $m(D) = \inf\{m(C) \mid D \subset C^o\}$

In a compact space every regular content can be extended to an unique regular Borel measure ([H2], §§53,54).

If m and m' are two finite Borel measures on a topological space and every set with zero m-measure has also m'-measure zero the Radon-Nikodym theorem states that there is an m-integrable function f such that m' = fm. The theorem can clearly be extended to apply to measures which are constructed from finite measures with pairwise disjoint support.

Boolean algebras

A Boolean algebra is a distributive complemented lattice with maximal and minimal element. With each Boolean algebra $\mathfrak{A}$ we associate a compact totally disconnected topological space Ω in the following manner. We consider the trivial Boolean algebra $2 := \{0,1\}$ as a topological space with the discrete topology and define Ω as the set of all homomorphisms of Boolean algebras from $\mathfrak{A}$ in 2. Thus Ω is a closed subspace of the compact totally disconnected space $2^{\mathfrak{A}}$ and so also compact and totally disconnected. Ω is called the

Stonean space of the Boolean algebra $\mathfrak{A}$. The mapping $a \mapsto B_a := \{ f \mid f \in \Omega, f(a) = 1 \}$ is an isomorphism of Boolean algebras from $\mathfrak{A}$ to the Boolean algebra of clopen subsets of Ω. (See e. g. [H1]).

A Boolean algebra in which every subset has an infimum and a supremum is called complete. A Boolean algebra is complete if and only if its Stonean space is extremally disconnected.

Every Boolean algebra is ordered in a natural way by the order $a \leq b \Leftrightarrow a \wedge b = a$. If $(\mathfrak{A}_i)_{i \in I}$ is a family of Boolean algebras the cartesian product of the $\mathfrak{A}_i$'s can be made into a Boolean algebra by defining the lattice operations component-wise. This Boolean algebra is called the product of the Boolean algebras $\mathfrak{A}_i$ and is written $\prod_{i \in I} \mathfrak{A}_i$.

Chapter 1: L^p-projections

X is always a Banach space over the reals. We define certain subspaces of X and investigate some properties which they have. These subspaces will be considered in much more detail in the following chapters.

1.1 Definition: Let $1 \leq p \leq \infty$, $J \subset X$ a subspace, $E : X \to X$ a projection (that is E linear, $E^2 = E$).

(i) J is called L^p-summand, if there is a subspace $J^\perp$ such that algebraically $X = J \oplus J^\perp$, and for $x \in J$, $x^\perp \in J^\perp$ we always have $\|x+x^\perp\|^p = \|x\|^p + \|x^\perp\|^p$ (if $p = \infty$: $\|x+x^\perp\| = \max \{\|x\|, \|x^\perp\|\}$).

(ii) E is called L^p-projection, if for every $x \in X$

$\|x\|^p = \|Ex\|^p + \|x-Ex\|^p$ (if $p = \infty$: $\|x\| = \max \{\|Ex\|, \|x-Ex\|\}$)

1.2 Proposition:

(i) For any L^p-summand J the subspace $J^\perp$ in definition 1.1(i) is uniquely determined. We therefore call $J^\perp$ "the L^p-summand complementary to J" and write $X = J \oplus_p J^\perp$.

(ii) Let J be an L^p-summand and E be the projection onto J with respect to $X = J \oplus_p J^\perp$. Then E is an L^p-projection.

(iii) For any L^p-projection E the spaces range E and ker E are complementary L^p-summands, that is $X = \text{range } E \oplus_p \ker E$.

(iv) Every L^p-projection E is continuous with $\|E\| \leq 1$.

In particular, L^p-summands J are closed (since $J = \ker E^\perp$, where $E^\perp$ is the L^p-projection onto $J^\perp$).

(v) There is a one-to-one correspondence between the set of

L^p-summands and the set of L^p-projections.

Proof:

(i) Let J be an L^p-summand, such that $J_1^\perp$ and $J_2^\perp$ satisfy the conditions of definition 1.1(i). We will prove $J_1^\perp = J_2^\perp$.

Let $y \in J_1^\perp$. We have $y = x + x^\perp$ where $x \in J$, $x^\perp \in J_2^\perp$. For $p < \infty$ it follows that $\|y\|^p = \|x\|^p + \|x^\perp\|^p$. On the other hand, $\|x^\perp\|^p = \|x\|^p + \|y\|^p$ (because $x^\perp = -x + y$, $x \in J$, $y \in J_1^\perp$), hence $x = 0$ and $y = x^\perp \in J_2^\perp$.

If $p = \infty$, consider $y + ax$ ($= (a+1)x + x^\perp$) for $a > 0$. Condition 1.1(i) implies $\max\{\|(a+1)x\|, \|x^\perp\|\} = \|y+ax\| = \max\{\|ax\|, \|y\|\}$, so necessarily $x = 0$.

We have thus proved that $J_1^\perp \subset J_2^\perp$. The reverse inclusion follows by symmetry.

(ii), (iii), (iv), (v) are easily verified. □

Examples:

1) Let $1 \leqq p \leqq \infty$ and (S, Σ, μ) a measure space. In $X = L^p(S, \Sigma, \mu)$, every measurable subset $B \subset S$ defines an L^p-projection by $f \mapsto f\chi_B$. The measurability of B is not essential. It is sufficient that for $f \in X$ always $f\chi_B \in X$ (that means $B \cap D \in \Sigma$ for $D \in \Sigma$, $\mu(D) < \infty$). We investigate the structure of the L^p-projections on X in more detail in chapter 4.

2) Every closed subspace J of a Hilbert space is an L^2-summand. $J^\perp$ is the usual space orthogonal to J, and the norm condition is the Pythagorean law for orthogonal elements.

3) Let T be a topological space and $S \subset T$ a clopen subset. The annihilator of S, $\{f \mid f : T \to \mathbb{R}$ continuous and bounded, $f|_S = 0\}$ is an L^∞-summand in the space of all real-valued continuous and bounded functions on T. We will show in appendix 2 that for compact T all L^∞-summands have this form.

4) The operators Id and 0 are always L^p-projections. We say that <u>the L^p-structure of X is trivial</u> if there are no other L^p-projections (or equivalently: there are no other L^p-summands than X and $\{0\}$).

5) If X and Y are Banach spaces, $1 \leq p \leq \infty$, define the norm on $X \times Y$ by $\|(x,y)\| := (\|x\|^p + \|y\|^p)^{1/p}$ (if $p = \infty$: $\|(x,y)\| = \max\{\|x\|, \|y\|\}$). As subspaces of $X \times Y$, X and Y are complementary L^p-summands, and up to isometric isomorphism L^p-summands always have this form.

We now state a theorem concerning L^p-projections which is fundamental to the following investigations. Motivated by results of [C] and [AE] (L^1- and L^∞-projections there are called L- and M-projections, respectively) it would seem reasonable to attempt to prove a commutativity theorem for L^p-projections which seemed to be essential for nearly all results. A thorough study of certain classes of Banach spaces (CK-spaces, AK-spaces, L^p-spaces; cf. [B1] and [Sü 2]) showed that in these classes L^p-projections always commute if $p \neq 2$, and every Banach space admits nontrivial L^p-projections for at most one p in $[1,\infty]$. Of course, L^2-projections will not commute in general, because every orthogonal projection on a Hilbert space is an L^2-projection.

<u>1.3 Theorem:</u> Let X be a Banach space over the reals.

(i) For every $p \in [1,\infty]$, $p \neq 2$, L^p-projections on X commute.

(ii) The space $(\mathbb{R}^2, \| \|_1)$ $(\cong (\mathbb{R}^2, \| \|_\infty))$ admits nontrivial L^1-projections and nontrivial L^∞-projections, as is easy to see. This space is the only Banach space which admits nontrivial L^p-projections for two different values of p. Equivalently: If X is not isometrically isomorphic to $(\mathbb{R}^2, \| \|_1)$, then nontrivial L^p-projections exist for at most one p in $[1,\infty]$.

<u>Proof:</u> The proof is elementary, but very involved. An outline of the proof which is published in [B2] is included in appendix 2. Here we only prove the following lemma which is essential for showing 1.3. It states that transposes of L^p-projections behave as expected.

<u>1.4 Lemma:</u> Let X be a Banach space, $1 \leq p \leq \infty$, $E : X \to X$ an L^p-projection. If we define $p' \in [1,\infty]$ by $1/p+1/p'=1$, then $E' : X' \to X'$ is an $L^{p'}$-projection. In particular, annihilators of L^p-summands of X in X' are $L^{p'}$-summands.

Conversely, if $E : X \to X$ is a continuous linear projection such that E' is an $L^{p'}$-projection, then E is an L^p-projection.

<u>Proof:</u> The second assertion is a consequence of the first, because E is obtained from E" by restriction and $p'' = p$. E' is a projection, so we only have to verify the norm condition. Because of $\|f \circ E\| = \|f|_{\text{range } E}\|$, $\|f \circ (Id-E)\| = \|f|_{\ker E}\|$ for $f \in X'$ we only have to show that $\|f\|^{p'} = \|f|_{\text{range } E}\|^{p'} + \|f|_{\ker E}\|^{p'}$

(for $p = 1$: $\|f\| = \max \{\|f|_{\text{range } E}\|, \|f|_{\ker E}\|\}$). If $p = 1$ or $p = \infty$ this is easily checked. For $1 < p < \infty$ it is a consequence of the equation $a^{p'} + b^{p'} = \max\limits_{0 \leq t \leq 1} (at + b(1-t^p)^{\frac{1}{p}})^{p'}$ (for $a,b \geq 0$), which can be obtained by elementary analytical techniques. □

Remark: The commutativity of L^1-projections can be shown with much less effort than in the general case. (Let E and F be L^1-projections on X, $x \in X$. We have $\|E\|, \|F\|, \|Id-E\|, \|Id-F\| \leq 1$ and therefore $\|Ex\| + \|x-Ex\| = \|x\| = \|Fx\| + \|x-Fx\| = \|EFx\| + \|Fx-EFx\| + \|x-Fx\| \geq$ $\geq \|EFx\| + \|Fx-EFx\| + \|Ex-EFx\| \geq \|Ex\| + \|Fx-EFx\|$ and $\|Ex\| + \|x-Ex\| =$ $= \|x\| = \|Fx\| + \|x-Fx\| = \|EFx\| + \|Fx-EFx\| + \|x-Fx\| \geq \|EFx\| + \|Fx-EFx\|$ $+ \|(Id-E)(x-Fx)\| \geq \|EFx\| + \|x-Ex\|$, that is $\|x-Ex\| \geq \|Fx-EFx\|$ and $\|Ex\| \geq \|EFx\|$. Consequently, $EFx = FEx$ if $Ex = 0$ or $Ex = x$, that means $EFE = FE$ and $EF(Id-E) = FE(Id-E)$. $EF = FE$ is now obtained by addition. By Lemma 1.4, L^∞-projections also commute.)

As a consequence of 1.3 it is now easy to prove some elementary facts about L^p-projections:

1.5 Proposition: Let X be a Banach space over $\mathbb{R}$, $p \in [1,\infty]$, $p \neq 2$, $\mathbb{P}_p$ the set of all L^p-projections on X ($\mathbb{P}_p(X)$ if we need to specify the Banach space).

(i) $E,F \in \mathbb{P}_p$ implies $E \circ F \in \mathbb{P}_p$. In particular, the intersection of two L^p-summands is again an L^p-summand (note range $E \circ F =$ range $E \cap$ range F)

(ii) $E,F \in \mathbb{P}_p$ implies $E + F - E \circ F \in \mathbb{P}_p$. As a consequence of the identity range $(E + F - E \circ F) =$ range $E +$ range F it follows that the sum of two L^p-summands is again an L^p-summand.

Therefore $\mathbb{P}_p$ is a Boolean algebra, if we define $E \wedge F = E \circ F$, $E \vee F = E + F - E \circ F$, $\overline{E} = Id - E$ (in particular: $E \leq F \Leftrightarrow EF = E$).

Proof:

(i) $E \circ F$ is a projection, because E and F commute. If $p < \infty$, we have for every $x \in X$

$$\|Ex\|^p = \|FEx\|^p + \|Ex-FEx\|^p \qquad (F \in \mathbb{P}_p)$$

$$\|x\|^p = \|Ex\|^p + \|x-Ex\|^p \qquad (E \in \mathbb{P}_p)$$

$$\|x-EFx\|^p = \|Ex-EFx\|^p + \|x-Ex\|^p \qquad (E \in \mathbb{P}_p)$$

which implies $\|x\|^p = \|EFx\|^p + \|x-EFx\|^p$.

The case $p = \infty$ is reduced to the foregoing one by transposition: E', F' are L^1-projections. Thus $E'F'$ $(=(EF)')$ is also an L^1-projection, so that, by 1.4, EF must be an L^∞-projection.

(ii) It is easy to check that $E + F - E \circ F$ is a projection. For $p < \infty$ and $x \in X$ we have

$$\|x\|^p = \|Ex\|^p + \|x-Ex\|^p \qquad (E \in \mathbb{P}_p)$$

$$\|x-Ex\|^p = \|Fx-FEx\|^p + \|x-Ex-Fx+FEx\|^p \qquad (F \in \mathbb{P}_p)$$

$$\|Ex+Fx-EFx\|^p = \|Ex\|^p + \|Fx-EFx\|^p \qquad (E \in \mathbb{P}_p)$$

and therefore $\|x\|^p = \|(E+F-EF)x\|^p + \|x-(E+F-EF)x\|^p$.

The case $p = \infty$ is reduced to the study of L^1-projections as in part (i). □

We now prove that for $p < \infty$, $p \neq 2$, $\mathbb{P}_p$ is in fact a complete Boolean algebra in which increasing (resp. decreasing) nets converge pointwise to their supremum (resp. infimum). This is not true for $p = \infty$, because even in CK-spaces the closure of the union of an ascending sequence of L^∞-summands is in general not an L^∞-summand.

Take for instance $\{(x_k) \mid k > n \Rightarrow x_k = 0\}$ for $n \in \mathbb{N}$, an ascending sequence of L^∞-summands in the space c of all convergent sequences. The closure of the union, that is the space of null sequences, is not an L^∞-summand, as is easily seen by using the result of appendix 2 ($\{\infty\}$ is not clopen in $\alpha\mathbb{N}$, the one-point compactification of the natural numbers, and $c = C(\alpha\mathbb{N})$). In the same way we can obtain more complicated counter-examples: If $T = \{0\} \cup \{\pm\frac{1}{n} \mid n \in \mathbb{N}\}$ (relative topology of the reals), $A_n := \{f \mid f \in CT,\ f(t) = 0 \text{ for } t \neq 1/n\}$, $B_n := \{f \mid f \in CT,\ f(t) = 0 \text{ for } t > 1/n\}$, then A_n and B_n are L^∞-summands in CT, but the A_n (resp. the B_n) have no supremum (resp. no infimum) in the set of all L^∞-summands in CT.

__1.6 Proposition:__ Let X be a Banach space, $1 \leq p < \infty$, $p \neq 2$.

(i) If $(E_i)_{i \in I}$ is an increasing net of L^p-projections, then $(E_i)_{i \in I}$ is pointwise convergent to an L^p-projection E. In $\mathbb{P}_p$ we have $E = \sup\{E_i \mid i \in I\}$.

(ii) Every downward filtrating net $(E_i)_{i \in I}$ is pointwise convergent to an L^p-projection E with $E = \inf\{E_i \mid i \in I\}$.

(iii) $\mathbb{P}_p$ is a complete Boolean algebra.

(iv) For any system $(J_i)_{i \in I}$ of L^p-summands the subspaces $J_1 := \bigcap_{i \in I} J_i$ and $J_2 := (\mathrm{lin}(\bigcup_{i \in I} J_i))^-$ are also L^p-summands. The respective L^p-projections are given by $\inf\{E_i \mid i \in I\}$ and $\sup\{E_i \mid i \in I\}$, (E_i = L^p-projection onto E_i).

__Proof:__ (i) For $x \in X$ and $i \geq j$ we have $\|x\| \geq \|E_i x\| \geq \|E_j x\|$. Therefore the net $(\|E_i x\|)_{i \in I}$ is increasing and bounded from above, so that its limit exists in $\mathbb{R}$. Also, the identity

$\|x\|^p = \|E_i x\|^p + \|x-E_i x\|^p$ implies the existence of the limit $\lim_{i\in I} \|x-E_i x\|$. Further, for $i \geq j$, $\|E_i x-E_j x\|^p = \|E_i x\|^p - \|E_j x\|^p$, hence $(E_i x)_{i\in I}$ is a Cauchy net in X. We define $E : X \to X$ by $Ex := \lim_{i\in I} E_i x$. It is obvious that E is linear and satisfies $\|x\|^p = \|Ex\|^p + \|x-Ex\|^p$ for all $x \in X$(letting i tend to infinity in $\|x\|^p = \|E_i x\|^p + \|x-E_i x\|^p$). We thus merely have to prove $E^2 = E$. To this end, let $x \in X$ and $\varepsilon > 0$. Choose $j_0 \in I$, such that $j \geq j_0$ implies $\|E_j Ex-E^2 x\| \leq \varepsilon$. For sufficiently large $k \in I$ (w.l.o.g. $k \geq j_0$) we have further that $\|E_k x-Ex\| \leq \varepsilon$ and therefore that $\|E_k x-E_k Ex\| \leq \varepsilon$. Hence $\|Ex-E^2 x\| \leq 3\varepsilon$, and thus $E = E^2$. Finally, the pointwise convergence implies $E_i E = EE_i = E_i$ for all $i \in I$ and, F being in $\mathbb{P}_p$, $E_i F = FE_i = F$ (all $i \in I$) $\Rightarrow FE = EF = E$. But that means $E = \sup \{E_i \mid i \in I\}$ (in $\mathbb{P}_p$).

(ii) is reduced to (i) by considering the increasing net $(Id-E_i)_{i\in I}$.

(iii) Since we have already proved that $\mathbb{P}_p$ is a lattice, the completeness is easily established by making use of (i) and (ii) with the standard techniques of lattice theory.

(iv) We first show that J_1 is an L^p-summand and $(\inf E_i)(X) = J_1$. As we already proved that $\inf E_i$ is an L^p-projection we only have to show that $(\inf E_i)(X) = \cap J_i$. W.l.o.g. we assume the E_i to be downward filtrating, such that $(\inf E_i)(x) = \lim E_i x$ for every $x \in X$. For $x \in \cap J_i$ we have $E_i x = x$ (all $i \in I$) and therefore $(\inf E_i)(x) = x$ which proves "$\supset$". Conversely, let $z = \lim E_i x \in (\inf E_i)(X)$. For every $j \in I$, $E_j z = \lim E_i E_j x = \lim E_i x$ (because for $i \leq j$, $E_i E_j x = E_i x$), so that $z = E_j z$ and therefore $z \in \cap J_i$. For the second part, we only have to prove $J_2 = (\sup E_i)(X)$, where w.l.o.g. the E_i's are upward filtrating, such that $J_2 = (\bigcup J_i)^-$ and $(\sup E_i)(x) = \lim E_i x$ for every $x \in X$. For $x \in X$ and $i \in I$, we have $E_i x \in J_i$, so $\lim E_i x \in (\bigcup J_i)^-$. To prove the reverse inclusion, first note that $E_j x = E_i E_j x$ for $i \geq j$ and therefore $J_i \subset (\sup E_i)(X)$. Thus $\bigcup J_i \subset (\sup E_i)(X)$, so that also $J_2 \subset (\sup E_i)(X)$, because the range of an L^p-projection is closed.

Finally we note that the identity $Id - \sup E_i = \inf (Id - E_i)$ implies $X = (\mathrm{lin} \bigcup J_i)^- \oplus_p \bigcap J_i$. □

<u>Remark:</u> Let $P[X]$ be the set of all continuous projections on X with norm ≤ 1, ordered by $P \leq Q \underset{\text{def.}}{\Leftrightarrow} PQ = QP = P$ (this order obviously extends the order on $\mathbb{P}_p$). It is easy to see that the sup (resp. the inf) of a finite family in $\mathbb{P}_p$ is also the sup (resp. the inf) in $P[X]$. The pointwise convergence implies that this is also true of arbitrary families.

The following proposition is devoted to the investigation of the heredetary properties of L^p-summands (for $p = 1$, cf. [AE]).

<u>1.7 Proposition:</u> Let J be an L^p-summand in X with associated L^p-projection E ($1 \leq p \leq \infty$, $p \neq 2$).

(i) The L^p-summands of J are exactly the L^p-summands of X which are contained in J.

(ii) The canonical mapping $\nu : X \to X/J$ induces an isometry between $J^\perp$ and X/J.

(iii) Images and inverse images of L^p-summands with respect to ν are also L^p-summands.

Proof:

(i) For any L^p-summand J_1 of J with projection E_1 the mapping E_1E is an L^p-projection from X onto J_1. Consequently, J_1 is an L^p-summand in X. Conversely, if $J_1 \subset J$ is an L^p-summand in X with projection E_1, then $E_1E = EE_1$ implies $E_1(J) \subset J$, so that $E_{1|J}$ is an L^p-projection with range J_1.

(ii) For $x \in \ker E$ and $y \in \text{range } E$ we have $\|x\| \leq \|x+y\|$ and therefore $\|[x]\| = \|x\|$. That $\nu \mid_{\ker E}$ is onto is obvious.

(iii) Let J_1 be an L^p-summand in X. Then $\nu(J_1) =$ $= \nu \mid_{\ker E} ((\ker E) \cap J_1)$ is, as the image of an L^p-summand w.r.t. a bijective isometric map, an L^p-summand in $\nu(X)$. Conversely, if $\mathfrak{J}$ is an L^p-summand in X/J, $J_1 := (\nu \mid_{\ker E})^{-1}(\mathfrak{J})$ is an L^p-summand in $\ker E$. Thus, $\nu^{-1}(\mathfrak{J}) = J_1 + \text{range } E$ is an L^p-summand in X ((i) and 1.5(i)). □

For L^p-summands, the essential condition is that the norm of an element is a special function of the norms of the components. More generally, one could consider the following definition.

Let $F : \mathbb{R}_+^2 \to \mathbb{R}_+$ be a fixed function. Two subspaces J, $J^\perp$ of X (X a real Banach space) are called complementary F-summands, if algebraically $X = J \oplus J^\perp$ and for $x \in J$, $x^\perp \in J^\perp$ $\|x+x^\perp\| = F(\|x\|, \|x^\perp\|)$. It is clear how F-projections have to be defined. In particular, L^p-summands are F-summands for $F = F_p$, where $F_p(s,t) := (s^p+t^p)^{1/p}$ if $p < \infty$ and $F_\infty = \max\{s,t\}$.

That there is nevertheless no restriction in only discussing L^p-summands follows from the following result due to R. Evans. If there is a Boolean algebra of F-projections which contains more than four elements (that means if there are at least three pairwise commuting nontrivial F-projections on some space), then necessarily $F = F_p$ for some p, $1 \leq p \leq \infty$. The idea of the proof is to show that F satisfies a functional equation which by a theorem of Bohnenblust has exactly the F_p as solutions. We note that in the theory of Banach lattices there is a similar result which is also proved by applying Bohnenblust's theorem: L^p-spaces and M-spaces are the only Banach lattices of dimension greater than two where the norm of the sum of disjoint elements is a function of the norm of the components (see [L], th. 15.5). We are therefore justified in restricting ourselves to the F_p's, that is to L^p-summands.

Chapter 2: The Cunningham p-Algebra

In this chapter we investigate the Banach algebra of operators generated by the set of all L^p-projections. For $1 \leq p \leq \infty$, $p \neq 2$, $C_p(X)$ denotes the uniform closure of the linear hull of $\mathbb{P}_p$:

$$C_p(X) := (\text{lin } (\mathbb{P}_p))^-.$$

$C_p(X)$ is a closed commutative subalgebra of the space of all bounded linear operators on X. It is called the Cunningham p-algebra of X (for p = 1 Cunningham algebra, see also [AE]).

Of fundamental importance for the theory is the representation of $C_p(X)$ as a space of continuous functions on a suitable compact topological space. This result (prop. 2.1) generalizes the one for p = 1 by Alfsen-Effros.

If Ω_p denotes the Stonean space of the Boolean algebra $\mathbb{P}_p$ ($1 \leq p \leq \infty$, $p \neq 2$), then we have:

2.1 Proposition: There is an isometric algebra-isomorphism of $C_p(X)$ onto $C(\Omega_p)$

Proof: (see Lemma 7.1 in [C1])

Since the family $\{\chi_{B_E} \mid E \in \mathbb{P}_p\}$ separates the points of Ω_p, we have $C(\Omega_p) = (\text{lin } \{\chi_{B_E} \mid E \in \mathbb{P}_p\})^-$. Therefore it is sufficient to construct an isometric algebra-isomorphism of $\text{lin}(\mathbb{P}_p)$ onto $\text{lin } \{\chi_{B_E} \mid E \in \mathbb{P}_p\}$. We call $\sum_{i=1}^{n} a_i E_i \in \text{lin}(\mathbb{P}_p)$ a canonical representation of the operator $T \in \text{lin}(\mathbb{P}_p)$, $T \neq 0$, if the E_i are non-zero and pairwise orthogonal (i. e. $E_i E_j = \delta_{ij} E_i$) and the a_i are all non-zero and distinct, and if $T = \sum_{i=1}^{n} a_i E_i$. By induction we obtain: Every $T \in \text{lin}(\mathbb{P}_p)$, $T \neq 0$, has exactly one canonical representation. Thus the following mapping Ψ of $\mathrm{lin}(\mathbb{P}_p)$ into $\mathrm{lin}\{\chi_{B_E} \mid E \in \mathbb{P}_p\}$ is welldefined: $T \mapsto \Psi(T) := \sum_{i=1}^{n} a_i \chi_{B_{E_i}}$ $(0 \neq T = \sum_{i=1}^{n} a_i E_i$ in canonical representation). Finally we define $\Psi(0) = 0$.

An easy computation shows that Ψ is linear. Ψ is isometric: Since the E_i are orthogonal, the canonical representation $T = \sum_{i=1}^{n} a_i E_i$ yields $\|\Psi(T)\| = \max\{|a_i| \ \ i=1,\ldots,n\}$ $(= |a_1|$ w. l. o. g.). Since $E_1 \neq 0$, there exists an element $x \in \mathrm{range}\ E_1$, $x \neq 0$. Hence $Tx = a_1 x$ and therefore $\|T\| \geq |a_1|$. On the other hand for arbitrary $x \in X$ we get for $1 \leq p < \infty$ $\|Tx\|^p = \|\sum_{i=1}^{n} a_i E_i x\|^p = \sum_{i=1}^{n} |a_i|^p \|E_i x\|^p \leq$ $|a_1|^p \|\sum_{i=1}^{n} E_i x\|^p \leq |a_1|^p \|\sum_{i=1}^{n} E_i\|^p \|x\|^p \leq |a_1|^p \|x\|^p$ and for $p = \infty$ $\|Tx\| = \|\sum_{i=1}^{n} a_i E_i x\| = \max\{|a_i| \|E_i x\| \ \ i=1,\ldots,n\} \leq |a_1| \|\sum_{i=1}^{n} E_i x\|$ $\leq |a_1| \|\sum_{i=1}^{n} E_i\| \ \|x\| \leq |a_1| \|x\|$. It follows that $\|T\| \leq |a_1|$ and thus $\|T\| = \|\Psi(T)\|$. Finally Ψ is onto, since every $f \in \mathrm{lin}\{\chi_{B_E} \mid E \in \mathbb{P}_p\}$ has a canonical representation as in the above definition, which generates an inverse image under Ψ in a natural way.

Because of the fact that $\Psi(EF) = \chi_{B_{EF}} = \chi_{B_E \cap B_F} = \chi_{B_E} \chi_{B_F} = \Psi(E)\Psi(F)$, $\Psi(\mathrm{Id}) = 1$ and the linearity of Ψ, $\mathrm{lin}(\mathbb{P}_p)$ and $\mathrm{lin}\{\chi_{B_E} \mid E \in \mathbb{P}_p\}$ are also isomorphic as algebras. The extension of Ψ to the closures will also be denoted by Ψ. □

In the following we use $T_f := \Psi^{-1}(f)$ for $f \in C(\Omega_p)$ and $\hat{T} := \Psi(T)$ for $T \in C_p(X)$.

<u>Note</u>: Since the proof only uses the fact that the Boolean algebra $\mathbb{P}_p$ consists of L^p-projections, we obtain an analogous result for an arbitrary Boolean algebra P consisting of L^p-projections $(1 \leq p \leq \infty$,

including $p = 2$), namely that $C := (\,\mathrm{lin}(P)\,)^-$ has a natural representation as $C(\,\Omega(P)\,)$, where $\Omega(P)$ is the Stonean space of P.

<u>2.2 Corollary</u>: (cf. [Ba1] 2.8) An operator $T \in C_p(X)$ is in $\mathbb{P}_p$ iff it is idempotent

<u>Proof</u>: From $T^2 = T$ we obtain $(\Psi(T))^2 = \Psi(T)$ and thus $\Psi(T) = \chi_A$ for a suitable clopen set A in Ω_p. From $A = B_E$, $E \in \mathbb{P}_p$, we have $\Psi(T) = \chi_{B_E} = \Psi(E)$ and hence $T = E$. □

The isomorphism of proposition 2.1 yields an order-structure on $C_p(X)$ in the following way: Let $C(\,\Omega_p)^+$ denote the cone of pointwise order on $C(\,\Omega_p)$. Then we define a cone $C_p(X)^+$ in $C_p(X)$ by

$$T \in C_p(X)^+ \quad \Leftrightarrow \quad \Psi(T) \in C(\,\Omega_p)^+ .$$

The induced order $\leq_{C_p(X)^+}$ coincides on $\mathbb{P}_p$ with the order $<_{\mathbb{P}_p}$, which $\mathbb{P}_p$ has as a Boolean algebra: For $E, F \in \mathbb{P}_p$ we have $E \leq_{C_p(X)^+} F \Leftrightarrow F - E \in C_p(X)^+ \Leftrightarrow \Psi(F - E) \in C(\,\Omega_p)^+ \Leftrightarrow \chi_{B_E} \leq \chi_{B_F} \Leftrightarrow B_E \cap B_F = B_E \Leftrightarrow B_{E \wedge F} = B_E \Leftrightarrow E \wedge F = E \Leftrightarrow E <_{\mathbb{P}_p} F$.

Since $C(\,\Omega_p)^+$ is closed we get the following characterization of the cone $C_p(X)^+$:

<u>2.3 Proposition</u>: An element of $C_p(X)$ is positive iff it can be approximated uniformly by nonnegative linear combinations of elements of $\mathbb{P}_p$. □

As far as the topological properties of Ω_p go, we note the following: By construction Ω_p is a compact, totally disconnected topological space ($1 \leq p \leq \infty$, $p \neq 2$). Since the Boolean algebra $\mathbb{P}_p$ is complete for $1 \leq p < \infty$, $p \neq 2$ (prop. 1.6 (iii)), Ω_p is extremally disconnected (chapter 0). By 3.2 Ω_p is in fact hyperstonean for these p (in the sense of [P] p. 144). The following result shows that we get every hyperstonean space in this way.

<u>2.4 Proposition</u>: Let $1 \leq p < \infty$, $p \neq 2$ and let Ω be a compact hyperstonean topological space. Then there is a Banach space X with the following property: The Stonean space Ω_p corresponding to $\mathbb{P}_p(X)$ is homeomorphic to Ω.

<u>Proof</u>: By [P] the space $C(\Omega)$ has a predual, which is an $L^1(\mu)$ by [G] : $L^1(\mu)' \cong C(\Omega)$. We set $X := L^p(\mu)$. By 4.10 we have $C_p(L^p(\mu)) \cong L^1(\mu)'$. Using proposition 2.1 we get the following chain of isometric isomorphisms: $C(\Omega_p) \cong C_p(L^p(\mu)) \cong L^1(\mu)' \cong C(\Omega)$. Hence Ω and Ω_p are homeomorphic. □

Since Ω_p is extremally disconnected for $1 \leq p < \infty, p \neq 2$, $C(\Omega_p)$ is a complete vector lattice. In particular we have (see also 1.6):

<u>2.5 Proposition</u>: For $1 \leq p < \infty$, $p \neq 2$, every increasing bounded net in $C_p(X)$ has a supremum in $C_p(X)$. □

<u>Remark</u>: Proposition 2.5 holds for $p = \infty$ iff Ω_∞ is extremally disconnected, i.e. iff the Boolean algebra $\mathbb{P}_\infty$ is complete.

As a first application of proposition 2.1 we determine the Cunningham p-algebra of p-products ($1 \leq p \leq \infty$, $p \neq 2$). By definition the p-product of a family of Banach spaces, $(X_i)_{i\in I}$, is for $1 \leq p < \infty$ the Banach space $\prod_{i\in I}{}^p X_i := \{ (x_i)_{i\in I} \mid (x_i)_{i\in I} \in \prod_{i\in I} X_i, \| (x_i)_{i\in I}\| := (\sum_{i\in I} \| x_i\|^p)^{1/p} < \infty \}$ and for $p = \infty$ $\prod_{i\in I}{}^\infty X_i := \{ (x_i)_{i\in I} \mid (x_i)_{i\in I} \in$

$\prod_{i\in I} X_i$, $\|(x_i)_{i\in I}\| := \sup_{i\in I} \|x_i\| < \infty \}$.

We first prove that the only way to get L^p-projections on the p-product is to build them up in a natural way from L^p-projections on the components.

<u>2.6 Proposition</u>: Let $(X_i)_{i\in I}$ be a family of Banach spaces and $1\leq p\leq\infty$, $p \neq 2$. Then $\mathbb{P}_p(\prod_{i\in I}{}^p X_i) = \prod_{i\in I} \mathbb{P}_p(X_i)$.

<u>Proof</u>:

a) Let $(E_i)_{i\in I} \in \prod_{i\in I} \mathbb{P}_p(X_i)$, then $\prod_{i\in I} E_i$ ($\prod_{i\in I} E_i((x_i)_{i\in I}) :=$ $(E_i x_i)_{i\in I}$) is an L^p-projection on $\prod_{i\in I}{}^p X_i$.

b) For $E \in \mathbb{P}_p(\prod_{i\in I}{}^p X_i)$ $F_i := p_i E j_i$ is an L^p-projection on X_i (p_i denotes the i-th projection and j_i the i-th injection, $i\in I$): F_i is idempotent, since for every $x_i \in X_i$ we have $F_i^2(x_i) = p_i E j_i p_i E j_i(x_i)$ $= p_i E E_i E j_i(x_i) = p_i E^2 E_i j_i(x_i) = p_i E j_i(x_i) = F_i(x_i)$ (here $E_i := j_i p_i$ is an L^p-projection on $j_i(X_i) \subset \prod_{i\in I}{}^p X_i$). Since $E j_i(X_i) \subset j_i(X_i)$ we have $j_i F_i(x_i) = E j_i(x_i)$ and therefore for every $x_i \in X_i$ and $1\leq p<\infty$: $\|x_i\|^p = \|j_i(x_i)\|^p = \|E j_i(x_i)\|^p + \|j_i(x_i) - E j_i(x_i)\|^p = \|j_i F_i(x_i)\|^p$ $+ \|j_i(x_i) - j_i F_i(x_i)\|^p = \|F_i(x_i)\|^p + \|x_i - F_i(x_i)\|^p$ (similarly for $p = \infty$). Finally we show $\prod_{i\in I} F_i = E$, i.e. $(\prod_{i\in I} (p_i E j_i))((x_i)_{i\in I}) =$ $E((x_i)_{i\in I})$ for every $(x_i)_{i\in I} \in \prod_{i\in I}{}^p X_i$. This follows from $p_i E((x_i)_{i\in I})$ $= p_i E_i E((x_i)_{i\in I}) = p_i E E_i((x_i)_{i\in I}) = p_i E j_i(x_i)$. □

<u>2.7 Proposition</u>: For a family of Banach spaces $(X_i)_{i\in I}$ and $1\leq p\leq \infty$, $p \neq 2$, the spaces $C_p(\prod_{i\in I}^p X_i)$ and $\prod_{i\in I}^\infty C_p(X_i)$ are isometrically isomorphic.

<u>Proof</u>: The following technical statement can easily be verified: For a family of compact topological spaces $(K_i)_{i\in I}$ the spaces $C(\beta \dot{\bigcup}_{i\in I} K_i)$

and $\prod_{i\in I}^{\infty} C(K_i)$ are isometrically isomorphic. By [S] p. 280 we have the homeomorphism $\Omega_p(\prod_{i\in I} \mathbb{P}_p(X_i)) \cong \beta \dot{\bigcup_{i\in I}} \Omega_p(\mathbb{P}_p(X_i))$. Therefore it follows by 2.1 and 2.6 and the above remark that $C_p(\prod_{i\in I} {}^pX_i) \cong C(\Omega_p(\mathbb{P}_p(\prod_{i\in I} {}^pX_i))) \cong C(\Omega_p(\prod_{i\in I} \mathbb{P}_p(X_i))) \cong C(\beta \dot{\bigcup_{i\in I}} \Omega_p(\mathbb{P}_p(X_i))) \cong \prod_{i\in I}^{\infty} C(\Omega_p(\mathbb{P}_p(X_i))) \cong \prod_{i\in I}^{\infty} C_p(X_i)$. □

The Cunningham p-algebras of L^p-summands J and of quotient spaces X/J can be obtained in a natural way:

<u>2.8 Proposition</u>: Let $J \subset X$ be an L^p-summand with corresponding L^p-projection E and $1 \leq p \leq \infty$, $p \neq 2$.

Then (i) $C_p(J) \cong C_p(X)\cdot E$

(ii) $C_p(X/J) \cong C_p(X)\cdot(Id - E)$

<u>Proof</u>: (i) By $\Phi(T) := (TE)E$ for every $T \in C_p(J)$ we define an isometric isomorphism $\Phi : C_p(J) \to C_p(X)\cdot E$. If $F \in C_p(J)$ is an element of $\mathbb{P}_p(J)$, then $FE \in \mathbb{P}_p(X)$ and therefore $(FE)E \in C_p(X)\cdot E$. Because of $C_p(J) = (\text{lin}\mathbb{P}_p(J))^-$ Φ is welldefined and linear. It is isometric, because $\Phi(T)|_J = T$ and $\Phi(T)|_{J^\perp} = 0$. For $F \in \mathbb{P}_p(X)$ we have $F|_J \in \mathbb{P}_p(J)$ with $\Phi(F|_J) = (FE)E$. Hence Φ is onto, since $C_p(X) = (\text{lin}\mathbb{P}_p(X))^-$.
(ii) By (i) we get $C_p(X)\cdot(Id - E) \cong C_p(\text{range }(Id - E))$. Because of proposition 1.7 (ii) range $(Id - E)$ and X/J are isometric isomorphic. Therefore $C_p(X)\cdot(Id - E) \cong C_p(X/J)$. □

Finally we determine the Cunningham p'-algebra of dual spaces. In preparation we prove

<u>2.9 Lemma</u>: Every L^q-projection E on a Banach dual space X' is weak*-continuous $(1 < q \leq \infty)$.

Proof: The proof is similar to the one given by [CER] for $q = \infty$ (see also [F]). By [G] and the theorem of Krein-Šmulian it is sufficient to show that (range E) $\cap S_1^{X'}$ and (range(Id-E)) $\cap S_1^{X'}$ are weak*-closed. We show this for (range E) $\cap S_1^{X'}$. Let $(x_\alpha)_{\alpha\in I}$ be a convergent net in (range E) $\cap S_1^{X'}$ with x_α being weak*-convergent to $x = x_1 + x_2 \in$ (range E) $\oplus$ (range(Id-E)). Since the norm is lower semicontinuous in the weak*-topology, we get $\|x\| \le \liminf \|x_\alpha\| \le 1$ and therefore $x \in S_1^{X'}$. Consider the net $y_\alpha := x_\alpha - x_1$ in range E, which is weak*-convergent to x_2. Then we have for $h > 0$:
$(1/h)x_2 + y_\alpha \to (1/h)x_2 + x_2$ and therefore $\|(1 + 1/h)x_2\| \le \liminf \|(1/h)x_2 + y_\alpha\|$, thus $\|x_2\| + \|x_2\|/h \le \liminf \|x_2 + hy_\alpha\|/h$.
From $\|x_2 + hy_\alpha\| = (\|x_2 + hy_\alpha\|^q)^{1/q} = (\|x_2\|^q + h^q\|y_\alpha\|^q)^{1/q} \le (\|x_2\|^q + h^q(1+\|x_1\|)^q)^{1/q} = (\|x_2\|^q + h^qc^q)^{1/q}$ for $1 < q < \infty$ it follows that $\|x_2\| \le h^{-1}((\|x_2\|^q + h^qc^q)^{1/q} - \|x_2\|)$. Hence $\|x_2\| \le c \lim_{k\to 0} k^{-1}((\|x_2\|^q + k^q)^{1/q} - \|x_2\|) = 0$. Therefore $x_2 = 0$ and $x \in$ (range E).
For the case $q = \infty$ see [CER]. □

Note: The statement of lemma 2.9 is false for q=1 (e.g. X=C[0,1]).

2.10 Proposition: For a Banach space X the spaces $C_p(X)$ and $C_{p'}(X')$ are isometrically isomorphic ($1 \leqq p < \infty$, $1/p + 1/p' = 1$)

Proof: The mapping $T \mapsto T'$ is a linear isometry of [X] in [X']. By lemma 1.4 the range of its restriction to $C_p(X)$ is contained in $C_{p'}(X')$. Since $\mathbb{P}_{p'}(X')$ generates $C_{p'}(X')$ this mapping is onto by lemma 2.9. □

Chapter 3: The Integral Module Representation

In the foregoing chapters we have defined L^p-projections and examined the properties, both of individual projections and of systems of commuting projections. In particular, we have seen that the Cunningham p-algebra of a Banach space X has a natural representation as the space of continuous functions $C(\Omega)$ on the Stonean space Ω of the Boolean algebra of L^p-projections. In this chapter we shall work in a slightly more general setting. We assume that we are given a complete (in the projection sense) Boolean algebra $\mathfrak{A}$ consisting of L^p-projections ($p < \infty$), but not necessarily containing them all. Since the representation of the Cunningham p-algebra only used the fact that the projections were L^p-projections and not that it contains all of them we obtain the analogous result for our arbitrary algebra $\mathfrak{A}$, namely that $C := (\text{lin}\ \mathfrak{A})^-$ has a natural representation as the space of continuous functions on the Stonean space of $\mathfrak{A}$ which we denote by Ω. In this way we can also apply the theory to the case $p = 2$.

In this chapter we shall derive a representation of the Banach space X itself as a space of functions on Ω in such a way that the multiplication of these functions by continuous functions on Ω will correspond to the application of the corresponding operators on X. In other words, if Ψ is the natural representation of C as $C(\Omega)$ we shall derive a representation Φ of X as some space of functions on Ω, say Y, such that $(\Phi(Tx))(k) = ((\Psi T)(k))((\Phi x)(k))$ for all k in

Ω, T in C, x in X. It is clear that the functions in Y will not in general be scalar. We shall assume for the moment that at each point k the functions take values in some as yet undefined Banach space X_k. We shall see later how to obtain suitable spaces X_k.

A: Measure

In the representation Ψ the projections in $\mathfrak{U}$ correspond to the characteristic functions of the clopen sets in Ω. This means that if a representation Φ exists, the application of a projection in $\mathfrak{U}$ to an element of X corresponds to the restriction of the representing function in Y to the corresponding clopen set. Since the elements of $\mathfrak{U}$ are L^p-projections we must norm Y in such a way that these restrictions are also L^p-projections. The most natural way to do this is to norm Y as a vector-valued $L^p(m)$ for some measure m:

$$\|f\|_Y = \left(\int_\Omega \|f(k)\|_{X_k}^p \, dm\right)^{1/p} .$$

Unfortunately the derivation of the Stonean space of $\mathfrak{U}$ determines the topology of Ω but cannot be used to define a measure on it. However $\mathfrak{U}$ is not an abstract Boolean algebra but an algebra of projections on X and the elements of X can be used to define measures on Ω in the following way.

3.1 Construction: For D clopen let E_D be the corresponding projection in $\mathfrak{U}$. Let x be an element of X. For all closed sets B we define $$m_x(B) := \inf\{\|E_D x\|^p \mid D \text{ clopen}, D \supset B\} = \|E_{\text{int } B}x\|^p .$$
Int B is clopen because Ω is extremally disconnected and the two definitions are equivalent since $E_{\text{int } B}$ is the infimum of E_D, $D \supset B$

in $\mathfrak{A}$ and thus also limit in the strong topology of operators. The properties of m_x follow directly from the definition and the fact that the projections are all L^p-projections:

(i) $0 \leq m_x(B) < \infty$ for all closed sets B

(ii) $B \subset D \Rightarrow m_x(B) \leq m_x(D)$ for all closed sets B, D

(iii) $m_x(B \cup D) \leq m_x(B) + m_x(D)$ for all closed sets B, D

(iv) $m_x(B \cup D) = m_x(B) + m_x(D)$ for all disjoint closed sets B,D

(v) $m_x(B) = \inf \{ m_x(D) \mid B \subset \operatorname{int} D \}$ for all closed sets B .

A function on the closed sets with these properties is called a regular content (see 0.2) and is the restriction to the closed sets of an unique regular Borel measure which we also write m_x. Suppose B is a nowhere dense Borel set, then $\operatorname{int} B^- = \emptyset$. Since $E_\emptyset = 0$ it follows that $m_x(B) \leq m_x(B^-) = \| E_{\operatorname{int} B^-} x \|^p = 0$, i. e. m_x vanishes on nowhere dense sets and thus also on sets of first category in Ω.

Although this construction provides us with measures on Ω for every x in X it does not provide us with one global measure which is what we want in m. However the equality

$$\int_D \|f(k)\|^p_{X_k} \, dm = \|\chi_D f\|^p_Y = \|E_D x\|^p = m_x(D) \quad \text{(for clopen D whereby f in Y corresponds to } x \in X)$$

restricts our choice of m to those measures with respect to which all the m_x's have derivatives. The next lemma shows how to construct such a measure out of the m_x's themselves.

<u>3.2 Lemma</u>: With X and $\mathfrak{A}$ as above there is a Borel measure on Ω such that (i) Every non-empty clopen set has positive measure

(ii) Every nowhere dense set has zero measure

(iii) Every non-empty clopen set contains another with finite measure.

Note: Such measures are called perfect. An extremally disconnected space is called hyperstonean if and only if there is a perfect measure on it. This is equivalent to the normal definition involving normal measures as can easily be seen by working through this lemma with the normal measures instead of the m_x's. 3.4 then provides the other half of the equivalence.

Proof: Let $(x_\alpha)_{\alpha \in I}$ be a maximal family of elements of X such that $\alpha \neq \beta$ implies $\operatorname{supp} m_{x_\alpha} \cap \operatorname{supp} m_{x_\beta} = \emptyset$, chosen by Zorn's lemma. The formula $m(B) := \sum_I m_{x_\alpha}(B)$ for all Borel sets B defines a Borel measure m on Ω. m clearly has property (ii) since each m_x has it. Let D be a non-empty clopen set in Ω. Then there are two possibilities:

(a) $D \cap \operatorname{supp} m_{x_\alpha}$ is non-empty for some α.

In this case m(D) is positive and $D \cap \operatorname{supp} m_{x_\alpha}$ is a non-empty clopen set in D with finite measure.

(b) $D \cap \operatorname{supp} m_{x_\alpha}$ is empty for all α.

Then since D is non-empty there exists an x in X with $E_D x = x$ and $x \neq 0$. But $m_x(D) = \|E_D x\|^p = \|x\|^p = m_x(\Omega)$ i. e. $\operatorname{supp} m_x \subset D$. This contradicts the fact that $(x_\alpha)_{\alpha \in I}$ is maximal.

Since only (a) can occur we have that m has properties (i) - (iii). □

Because of property (i) all the m_x's are absolutely continuous with respect to any perfect measure on Ω. By (iii) any perfect measure can be decomposed into finite measures and thus the Radon-Nikodym theorem can be applied to obtain a derivative dm_x/dm. The perfect measures are thus the measures which we require for our representation.

B: Norm resolution

We assume now that we have chosen a perfect measure m on K, which is not necessarily constructed as in the lemma. We could now attempt to define Y and show that there is a mapping from X into Y with the desired properties. In order to do this however we should need to define the range spaces X_k. If k is an isolated point $\{k\}$ is clopen and corresponds to a projection E_k in $\mathfrak{A}$. It is natural in this case to define $X_k = E_k X$. On the other hand if k is not isolated there is no space which is natural in this way. We proceed therefore in the following manner. We are looking for a map from X into Y, which we assume to consist of vector-valued functions on Ω p-integrable with respect to m. If we had such a mapping we could go over to the associated scalar functions ($f \mapsto f^{\|\cdot\|}$ whereby $f^{\|\cdot\|}(k) = \|f(k)\|_{X_k}$) and thereby obtain a mapping from X into $(L^p(m))^+$. We shall denote this map by $[\cdot]$.

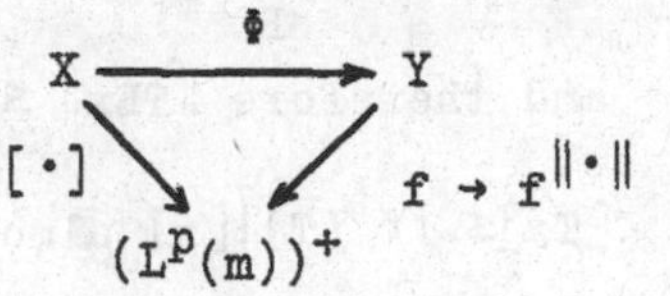

The map $[\cdot]$ must have the following three properties if it is to come from a representation of the desired form.

(1) $\|[x]\|_p = \|x\|$ for all x.

Since both Φ and $f \to f^{\|\cdot\|}$ are isometries

<u>3.3 Lemma</u>: With X, $\mathfrak{A}$, and m as above, there is a norm resolution for X taking values in $L^p(m)$.

<u>Proof</u>: We have seen how each element x of X defines a Borel measure on Ω, m_x, which has a derivative with respect to m, i. e. $m_x = f_x m$ for some $L^1(m)$-function f_x. Since m_x is positive f_x is positive almost everywhere so we can choose it to be positive everywhere. We define a mapping $[\cdot]$ from X into $(L^p(m))^+$ by $[x] := (f_x)^{1/p}$. $[\cdot]$ has the three defining properties of a norm resolution:

(i) $\|[x]\|_p^p = \int_\Omega f_x dm = m_x(\Omega) = \|x\|^p$

(ii) Suppose there is an element $T \in C$ and an element $x \in X$ for which $[Tx] \leq |\Psi(T)|[x]$ almost everywhere is false. Then there is a set B of positive m-measure such that $[Tx] \geq s$, $|\Psi(T)| \leq r$, $[x] \leq t$ on B whereby $s > rt$. Let D be the clopen set which differs from B by a set of first category in Ω and E the corresponding projection in $\mathfrak{A}$. $\|ETx\|^p = \int_B [Tx]^p dm \geq s^p m(B)$. $\|TEx\|^p \leq r^p \|Ex\|^p = \int_B r^p [x]^p dm \leq r^p t^p m(B)$ since $\|T|_{\text{range } E}\| = \|\Psi(T)\chi_B\| \leq r$ and therefore $\|TEx\| \leq r\|Ex\|$. However $ET = TE$. It follows that $[Tx] \leq |\Psi(T)|[x]$ almost everywhere. Similarly $[Tx] \geq |\Psi(T)|[x]$.

(iii) From (i) and (ii) follows, exactly as (3) followed from (1) and (2), that $[x+y] \leq [x] + [y]$ for all x, y in X. □

Note that if M is an arbitrary norm resolution for X taking values in $L^p(m)$ then if x is in X we have that for every clopen set D in K $m_x(D) = \|E_D x\|^p = \|M(E_D x)\|_p^p = \|\chi_D M(x)\|_p^p = \int_D M(x)^p dm$ (E_D in $\mathfrak{A}$ corresponds to D) so that $m_x = M(x)^p m$. Therefore the norm resolution of lemma 3.3 is unique.

C: The space $C^p(\Omega;m)$

In order to proceed to the construction of Y it is necessary to modify the norm resolution which we obtained in the last section. This may seem strange as we have just shown that this norm resolution is unique. However, the modification is essentially of a notational character so that there is no inconsistency in this. The main defect of our norm resolution is that its values are not functions; this fact was somewhat obscured because we have, up till now, made no distinction between p-integrable functions and the elements of $L^p(m)$ which are actually equivalence classes of functions. In order that $[\cdot]$ be the composition of Ψ with $f \to f^{\|\cdot\|}$ it is necessary that its values be actual functions. Since the norm resolution is unique this can only be done by choosing a function from each equivalence class in $L^p(m)$ and letting the resolution take this chosen function as its value instead of the equivalence class. We could of course choose the functions at random but it is more sensible and vital to the construction of Y that the functions chosen from equivalence classes which stand in some relation to each other, are themselves related in the same way. In particular, if for every p-integrable f the function chosen from the equivalence class containing f is f' we require:

(i) f ≧ 0 almost everywhere implies f' ≧ 0 everywhere.

(ii) f+g = h a.e. implies f'+g' = h' everywhere it is defined.

(2) $[Tx] = |\Psi(T)|[x]$ for all $T\in C$, $x\in X$.

It follows from the norm condition $\|ay\| = |a|\,\|y\|$ in any Banach space and in particular in every X_k that $(gf)^{\|\cdot\|} = |g|f^{\|\cdot\|}$ for every $g \in C(\Omega)$ and $f \in Y$. Since we require $\Phi(Tx) = \Psi(T)\Phi(x)$ we must have (2) for $[\cdot]$.

(3) $[x+y] \leq [x] + [y]$ for all x, y.

Suppose $[x+y] \not\leq [x] + [y]$ for some x, y in X. Then there is a set of positive m-measure on which $[x+y](k) \geq s$, $[x](k) \leq r$, $[y](k) \leq t$ with $s > r+t$. Let D be a clopen set which differs from this set, which we call B, by a set of first category in Ω and E the corresponding projection in $\mathfrak{A}$. By (2) $[E(x+y)] = \chi_D[x+y]$, $[Ex] = \chi_D[x]$ and $[Ey] = \chi_D[y]$ (since $\Psi(E) = \chi_D$). It follows that $\|[E(x+y)]\|_p^p = \int_D [x+y]^p dm = \int_B [x+y]^p dm \geq s^p m(B)$ (since $m(D\Delta B) = 0$). Similarly $\|[Ex]\|_p \leq r(m(B))^{1/p}$ and $\|[Ey]\|_p \leq t(m(B))^{1/p}$. So $\|[Ex]\|_p + \|[Ey]\|_p < \|[E(x+y)]\|_p$ in contradiction to (1).

If K is a compact topological space, V a Banach C(K)-module, Z a Banach function space on K and M a mapping from V into Z^+ such that

(i) $\|v\|_V = \|M(v)\|_Z$ for all v in V

(ii) $M(f\cdot v) = |f|\, M(v)$ for all f in C(K), v in V

(iii) $M(v+w) \leq M(v) + M(w)$ for all v, w in V

we say that M is a <u>norm resolution</u> for V taking values in Z. (1), (2) and (3) show that we are looking for a norm resolution for X taking values in $L^p(m)$. X is a $C(\Omega)$-module by virtue of the map Ψ.

(iii) $(rf)' = rf'$ for every scalar r.

(iv) $(|f|)' = |f'|$ for all f.

In general it is not possible, given a measure space (S,Σ,μ), to choose functions in the equivalence classes of $L^p(\mu)$ so that (i) - (iv) are satisfied. In our case, however, there is a particularly simple choice which satisfies these conditions.

<u>3.4 Lemma</u>: Let K be an extremally disconnected compact topological space and m a perfect measure on K. Then, every m-measurable function on K is equal almost everywhere to some continuous numerical function.

<u>Proof</u>: Consider the map $C(K) \to L^\infty(m)$ defined by $f \to$ (the equivalence class containing f). Then

a) This map is linear

b) If $|f(k)| > r$ for some k then since f is continuous there is an open set O with $k \in K$ on which $|f(l)| > r$. Since every non-empty open set has positive m-measure, the absolute supremum of f (the norm in $C(K)$) and the almost everywhere supremum (the norm in $L^\infty(m)$) are identical. i. e. the mapping is an isometry.

c) If $\sum_{i=1}^{n} r_i \chi_{B_i}$ is a measurable step function let D_i be the clopen set which differs from B_i by a set of first category in K ($i = 1,\dots,n$). Then $\sum_{i=1}^{n} r_i \chi_{D_i}$ is continuous and is mapped into the equivalence class containing $\sum_{i=1}^{n} r_i \chi_{B_i}$ (since $m(B_i \Delta D_i) = 0$ for all i). This means that the range of this mapping includes all step functions. Since the step functions are dense in $L^\infty(m)$ the mapping is onto.

For every numerical function f on K let f^+ be the function defined by $f^+(k) := f(k)/(1+|f(k)|)$ with $\pm\infty/(1+\infty) := \pm 1$ and for every function g from K into $[-1,+1]$ let g^- be the function $g^-(k) := g(k)/(1-|g(k)|)$ whereby $\pm 1/0 := \pm\infty$.

Suppose f is a numerical m-measurable function on K. Then f^+ is a bounded m-measurable function with norm ≤ 1. By c) there is a continuous function with norm ≤ 1 which is equal to f^+ almost everywhere, say g. g^- is then a continuous numerical function on K which differs from f on a set of zero m-measure. □

Since two different continuous functions differ on a non-empty open set which cannot be a set of zero m-measure the function in lemma 3.4 is unique. In particular every equivalence class in $L^p(m)$ contains exactly one continuous function. We call the model of $L^p(m)$ in which every equivalence class is represented by its unique continuous member <u>$C^p(K;m)$</u>. This model clearly has properties (i) - (iv).

We now make the norm resolution of the last section take values in $C^p(\Omega;m)$ instead of $L^p(m)$ i.e. $[x]$ denotes now, not an equivalence class, but the unique continuous member of this equivalence class. Apart from ensuring that the norm resolution now takes functions as values this has the important consequence that (ii) and (iii) of the definition of a norm resolution are now valid everywhere and not just almost everywhere. This follows from our observation above that two continuous functions which are equal almost everywhere are identical. In particular the evaluation of $[\cdot]$ at any point k has the properties:

a) $[x+y](k) \leq [x](k) + [y](k)$ for all x, y in X

b) $[rx](k) = |r|[x](k)$ for all r in $\mathbb{R}$, x in X (since $\Psi(rId)=r$).

Since the evaluation of $[\cdot]$ at any point k is a mapping from X into the positive extended real numbers, a) and b) show that it is a semi-norm, or rather would be if it were not for the fact that $[x](k)$ can take the value $+\infty$.

D: Direct Integrals and Integral Modules

Since we have been assuming that Y, the space we are looking for, is a sort of vector-valued L^p-space, it is now time to define such a space. If $[\cdot]$ is to be the composition of ϕ and $f \to f^{\|\cdot\|}$, we must have that $f^{\|\cdot\|} \in C^p(\Omega;m)$ for all $f \in Y$. Since the functions in $C^p(\Omega;m)^+$ take the value ∞ we must relax our assumption that the functions in Y take values in Banach spaces X_k in order to allow them to take infinite values as well. This leads to the following definition:

3.5 Definition: Let m be a perfect measure on an extremally disconnected compact topological space K and $(X_k)_{k\in K}$ a family of Banach spaces indexed by the points of K.

$$\int_K^{\oplus p} X_k dm := \{f \mid f:K \to \{\infty\} \cup \dot{\bigcup_{k\in K}} X_k;\ f(k) \in X_k \cup \{\infty\} \text{ for all } k;\ f^{\|\cdot\|} \in C^p(K;m)\}\ (\text{when } f(k)=\infty, \text{ put } f^{\|\cdot\|}(k)=+\infty)$$

is the p-direct integral of the Banach spaces X_k with respect to m.

For an element f of a p-direct integral we define $N(f) := \|f^{\|\cdot\|}\|_p$.

This is not a norm, indeed a p-direct integral is not in general a vector space under the pointwise operations.

Unless otherwise stated we shall restrict ourselves to direct integrals where the set of k for which $X_k = \{0\}$ is nowhere dense. Strictly speaking such direct integrals are called <u>essential</u>.

<u>Examples</u>: Let K be $\beta\mathbb{N}$ and $X_k \cong \mathbb{R}$ for all k. Let m be the measure which counts the points of $\mathbb{N}$. If f is an element of the p-direct integral of the X_k with respect to m, then integrability implies that $f(n) \to 0$ and thus continuity requires that $f(k) = 0$ for $k \notin \mathbb{N}$. The direct integral is then identical with $L^p(m)$. If however we give each point n of $\mathbb{N}$ the weight $1/2^n$ so that $m(K) = 1$, the functions in the direct integral no longer need to vanish on $\beta\mathbb{N} \setminus \mathbb{N}$. In this case both 1_K and $\chi_{\mathbb{N}} - \chi_{\beta\mathbb{N}\setminus\mathbb{N}}$ lie in the direct integral whereas their sum, $2\chi_{\mathbb{N}}$ does not since its absolute value is not continuous. In this case the direct integral is not a vector space.

Suppose Z is a subset of a p-direct integral over K with the following properties:

(i) for all f, g in Z there is an h in Z with $f(k) + g(k) = h(k)$ wherever it is defined

(ii) for all f in C(K), g in Z there is an h in Z with $f(k)g(k)=h(k)$ where it is defined.

Then if $h_1, h_2 \in Z$ and $h_1 = h_2$ almost everywhere, (i) and (ii) imply that there is a function $h_0 \in Z$ which is equal to $h_1 - h_2$ where it is defined. Since $h_1 - h_2$ is almost everywhere zero, h_0, and thus $h_0^{\|\cdot\|}$ is almost everywhere zero. As $h_0^{\|\cdot\|}$ is continuous h_0 is equal to zero everywhere. Therefore $h_1 = h_2$ wherever they are both defined. By the continuity of $h_1^{\|\cdot\|}$ and $h_2^{\|\cdot\|}$ and the fact that the set of points where h_1 and h_2 take the value ∞ is of first category h_1 and h_2 become infinite at the same points and thus they are equal everywhere. This shows that the relevant function h in (i) and (ii) is uniquely determined and thus that (i) and (ii) define an addition and multiplication for Z which clearly make it a C(K)-module. The restriction of N(·) to Z is a norm.

Such subsets exist in every direct integral. For example, if f is any element which does not take the value ∞ the set $\{gf \mid g \in C(K)\}$ has these two properties since $(g_1 + g_2)f = g_1f + g_2f$ and $g_1(g_2f) = (g_1g_2)f$ everywhere.

<u>3.6 Definition</u>: If in a direct integral there is a subset Z with (i) and (ii) and for which also

(iii) $X_k = \{f(k) \mid f \in Z,\ f(k) \neq \infty\}^-$ for all k

(iv) Z is complete in N(·)

we say that Z is a <u>p-integral module</u>.

A p-integral module is a Banach C(K)-module with the structure induced by (i) and (ii) (pointwise addition and multiplication almost everywhere) and the norm N(·). Unfortunately there is at present no way of telling given K, m, and the X_k, whether $\int_K^{\oplus p} X_k dm$ contains a p-integral module.

<u>E: The Representation</u>

Let us recap on the results of this chapter up till now. We started out from a Banach space X and a complete Boolean algebra of L^p-projections $\mathfrak{U}$ on it. We had already seen in the previous chapter that $C := (\text{lin } \mathfrak{U})^-$ is isometrically isomorphic to $C(\Omega)$. In the foregoing sections we saw that there exist perfect measures on Ω and that if m is an arbitrary perfect measure there is a norm resolution for X taking values in $C^p(\Omega;m)$. In section D we defined the p-direct integral of a family of Banach spaces $(X_k)_{k \in K}$ over K with respect to m. We also defined the concept of a p-integral module. Suppose now that Z is an integral module in some p-direct integral over Ω with respect to m. Then Z is a $C(\Omega)$-module, as is X, and the mapping $f \to f^{\|\cdot\|}$ is a norm resolution for Z taking values in $C^p(\Omega,m)$. This leads us to the assumption that the undefined representing space Y is a p-direct integral module for some suitable X_k.

We can now state and prove the representation theorem.

<u>3.7 Theorem</u> (The Integral Module Representation):

Let X, $\mathfrak{U}$, C, Ω, and m be as above.

Then there exist Banach spaces X_k, indexed by the points k of Ω and a p-integral module Y in the p-direct integral of the X_k with respect to m such that X and Y are isometrically isomorphic as Banach $C(\Omega)$-modules (X is a $C(\Omega)$-module by virtue of Ψ).

<u>Proof</u>: Let $[\cdot]$ be the (unique) norm resolution for X taking values in $C^p(\Omega;m)$. Then as noted at the end of section C, for every k

a) $[x + y](k) \leq [x](k) + [y](k)$ for all x,y in X

b) $[rx](k) = |r|[x](k)$ for all r in $\mathbb{R}$, x in X.

Consider for some k the set $Y_k := \{x \mid x \in X, [x](k) \neq \infty\} \subset X$. Then by a) and b), Y_k is closed under addition and scalar multiplication

i. e. is a subspace of X. Since $[\cdot](k)$ only takes finite values on Y_k it is a semi-norm for this space. Let X_k be the Banach space derived from Y_k with $[\cdot](k)$: $X_k := (Y_k/\{x \mid x \in X, [x](k) = 0\})^\wedge$. For convenience we write $\|\cdot\|_k$ for the norm on X_k induced by $[\cdot](k)$. For each element x of X define $\langle x \rangle$ by

$\langle x \rangle(k)$:= the equivalence class of x in X_k (if $x \in Y_k$)

$\langle x \rangle(k) := \infty$ (if $x \notin Y_k$).

Then $\langle x \rangle$ is a function on Ω which at each point k takes a value in $X_k \cup \{\infty\}$. Furthermore $\|\langle x \rangle(k)\|_k = [x](k)$ by definition so that $\langle x \rangle^{\|\cdot\|} = [x] \in C^p(\Omega;m)$. Thus $\langle x \rangle \in \int_\Omega^{\oplus p} X_k \, dm$ for every x in X. Let $Y := \{\langle x \rangle \mid x \in X\}$.

(i) Since $Y_k \to X_k$ is a homomorphism for every k, $\langle x \rangle(k) + \langle y \rangle(k) = \langle x+y \rangle(k)$ wherever it is defined (x, y in X)

(ii) Similarly $\langle rx \rangle(k) = r\langle x \rangle(k)$ wherever it is defined (r in $\mathbb{R}$, $x \in X$)

(iii) Suppose $g \in C(\Omega)$, $x \in X$. Let $T \in C$ be the operator corresponding to g. Then for each $k \in \Omega$, $[(T-g(k)Id)x] = |g-g(k)1_\Omega|[x]$ so that $[x](k) \neq \infty$ implies $[(T-g(k)Id)x](k) = 0$. From (i) and (ii) however follows that $\langle Tx \rangle - g(k)\langle x \rangle = \langle (T-g(k)Id)x \rangle$. Thus $[x](k) \neq \infty$ implies $\langle Tx \rangle(k) = g(k)\langle x \rangle(k)$. This is true for all k, so $\langle Tx \rangle = g\langle x \rangle$ where it is defined.

(iv) $N(\langle x \rangle) = \|\langle x \rangle^{\|\cdot\|}\|_p = \|[x]\|_p = \|x\|$

(v) From the construction $X_k = \{\langle x \rangle(k) \mid x \in X, \langle x \rangle(k) \neq \infty\}^-$.

Y is therefore a p-integral module, and the mapping $x \to \langle x \rangle$ is a $C(\Omega)$-module isomorphism and an isometry. □

Since the elements of an integral module are functions on the underlying space Ω and the isomorphism is one between $C(\Omega)$-modules, the representation whose existence we have just proved has the properties we were looking for. Although this representation is clearly not unique, since we had to choose a perfect measure on Ω, we shall see in the next section that the range spaces X_k which are non-trivial are uniquely determined by X and $\mathfrak{U}$.

F: Some properties of integral modules

Since we have just shown how to represent a Banach space as an integral module it is instructive to examine the properties of such modules. To this end we shall prove some fairly simple results concerning the structure of integral modules, one or two of which will come in useful later. The key step in most of these proofs is an application of Zorn's Lemma which in certain cases allows us to infer that a property of clopen sets holds for a certain set from the fact that it holds locally inside the set. In order not to have to repeat this step in each proof, we shall prove it separately.

Note: In this section we only consider abstract integral modules. Because of the representation theorem of the last section the results also apply to the general case. We write the scalar function associated with an element x either $\| x(.)\|$ or $[x]$. Since the norm resolution is unique, this can cause no confusion.

3.8 Existence Lemma:

Let $X \subset \int_K^p X_k \, dm$ be an integral module, B a non-void clopen subset of K with finite m-measure and A a subset of X such that the norm resolutions of all elements of A are bounded above by a common bound M. If A has the following properties:-

(i) for every clopen $C \subset B$, $C \neq \emptyset$ there is an $x \in A$, $x \neq 0$ with $\operatorname{supp} x \subset C$,

(ii) if x and y are in A and $\operatorname{supp} x \cap \operatorname{supp} y = \emptyset$ then $x + y$ is in A,

(iii) if $x \in X$, $C_1 \subset C_2 \subset \ldots C_n \subset \ldots$ a chain of clopen subsets of B, and $\chi_{C_n} x \in A$ for all n, then $\chi_C x \in A$ whereby $C = (\bigcup_{n=1}^{\infty} C_n)^{-}$,

then there exists an $x \in A$ with $\operatorname{supp} x = B$.

<u>Proof</u>: Let $A_o := \{ x \mid x \in A \text{ and } \operatorname{supp} x \subset B \}$. Order the elements of A_o by $x \leq y$ if and only if y is an extension of $x|_{\operatorname{supp} x}$. Let $\ldots \leq x_\alpha \leq \ldots \leq x_\beta \leq \ldots$ be an increasing chain in A_o. Then $\ldots N(x_\alpha) \leq \ldots \ldots \leq N(x_\beta) \leq \ldots\ldots M \cdot m(B)^{1/p}$. It follows from $N(x_\beta - x_\alpha)^p = N(x_\beta)^p - N(x_\alpha)^p$ for $x_\beta \geq x_\alpha$ that this chain is a Cauchy net in X. Let $x_1, x_2, x_3, \ldots$ be a cofinal subset. Let $x := \lim_n x_n$, $C_n := \operatorname{supp} x_n$ for each n and $C := (\bigcup_{n=1}^{\infty} C_n)^{-}$. Since $\chi_{C_n} x_m = x_n$ for all $m \geq n$, $\chi_{C_n} x = x_n$ for all n and since $\chi_C x_n = x_n$ for each n, $\chi_C x = x$. By (iii) it follows that $x \in A_o$ and $x \geq x_n$ for all n. By Zorn's Lemma A_o contains a maximal element and by (i) and (ii) the support of this element is B. □

As a first application of this lemma we show that the requirement that the evaluation of an integral module at a point be dense in the component space implies that an integral module is maximal as a set with the almost pointwise operations.

<u>3.9 Lemma</u>: An integral module $X \subset \int_K^p X_k \, dm$ is maximal amongst those subsets of the direct integral which are closed under almost pointwise addition.

Proof: Suppose Y is a subset of the direct integral, closed under almost pointwise addition, containing X and also some element y not lying in X. Let $\varepsilon > 0$, then there is a clopen set D in K such that a) $m(D) < \infty$, b) y is finite (and therefore bounded) on D, c) $N(\chi_{K\setminus D}y) < \frac{\varepsilon}{2}$. We apply the existence lemma to the set of all $x \in X$ with $\operatorname{supp} x \subset D$ and $\|x(k) - y(k)\| \leq \frac{\varepsilon}{2} m(D)^{-1/p}$ for $k \in \operatorname{supp} x$. (ii) is trivial. (iii) follows from the continuity of the norm resolution and (i) from the density of the x(k) in X_k. By the existence lemma there is an element $x \in X$ with $\operatorname{supp} x = D$ and $\|x(k) - y(k)\| \leq \frac{\varepsilon}{2} m(D)^{-1/p}$ for all $k \in D$. Then $N(y-x)^p = \int_K \|y(k) - x(k)\|^p \, dm =$

$$\int_{K\setminus D} \|y(k)\|^p \, dm + \int_D \|y(k) - x(k)\|^p \, dm < \frac{\varepsilon^p}{2^p} + \int_D \frac{\varepsilon^p}{2^p m(D)} \, dm = \frac{2\varepsilon^p}{2^p} \leq \varepsilon^p.$$

Since ε was arbitrary and X is complete, $y \in X$ which contradicts the assumption. □

The next lemma shows that the completeness of an integral module implies the completeness of the space of evaluations at each point, i.e. that each point of a component space is the evaluation of a suitable module element.

3.10 Lemma: Let $X \subset \int_K^p X_k \, dm$ be an integral module. For each $k \in K$ and every $x_k \in X_k$ there is an $x \in X$ such that $x(k) = x_k$.

Proof: We omit the elementary case where k is an atom. Thus we may assume that k has clopen neighbourhoods of arbitrary small measure. Since the values attained by the elements of X are dense in X_k, we can find elements $z_1, z_2, \ldots$ of X such that $\|z_n(k) - x_k\| < \frac{1}{2^n}$ for all $n \in \mathbb{N}$. Define a sequence (x_n) as follows. $x_1 = z_1$. Given x_{n-1} with $x_{n-1}(k) = z_{n-1}(k)$ let D_n be a clopen set such that a) $k \in D_n$,

b) $m(D_n) < \frac{1}{2^n}$, c) $\|z_n(l) - x_{n-1}(l)\| < \frac{1}{2^{n-3}}$ for all l in D_n. Since $\|z_n(k) - x_{n-1}(k)\| = \|z_n(k) - z_{n-1}(k)\| < \frac{1}{2^{n-2}}$ we can find such a set by continuity. Let $x_n = x_{n-1} + \chi_{D_n}(z_n - x_{n-1})$. Since $k \in D_n$, $x_n(k) = z_n(k)$. Clearly the sequence (x_n) is a Cauchy sequence in $\underset{\sim}{X}$ since $N(x_n - x_{n-1})^p = \int_{D_n} \|z_n(l) - x_{n-1}(l)\|^p \, dm < \frac{1}{2^n}(\frac{1}{2^{n-3}})^p$ and $p \geq 1$. Let x be the limit in X of this Cauchy sequence. Then x(k) must be x_k for if it were not there would be some n such that $\|x(k) - x_k\| > \frac{1}{2^n}$. But then $\|x(k) - x_{n+5}(k)\| > \frac{1}{2^{n+1}}$ so that there is some clopen set D containing k such that $\|x(l) - x_{n+5}(l)\| > \frac{1}{2^{n+2}}$ for l in D. Since D is non-empty there is an r such that $m(D) > \frac{1}{2^r}$, w.l.o.g. $r \geq n+5$. Let B be the clopen set $K \setminus (\bigcup_{i>r+1} D_i)^-$. Then $B \cap D$ is non-empty since $m(D) > \sum_{i>r+1} m(D_i)$. Now $\chi_B x_s = \chi_B x_r$ for $s \geq r$ so that $x = x_r$ on B and thus on at least one point of D. But $\|x_{n+5}(t) - x_r(t)\| \leq \|x_{n+5}(t) - x_{n+6}(t)\| + \ldots + \|x_{r-1}(t) - x_r(t)\| < \frac{1}{2^{n+3}} + \ldots + \frac{1}{2^{r-3}} < \frac{1}{2^{n+2}}$ for all $t \in K$. This contradicts the fact that $\|x(l) - x_{n+5}(l)\| > \frac{1}{2^{n+2}}$ for l in D. □

The following results are concerned with the relationship between an integral module and the particular perfect measure with respect to which the direct integral is taken. Since any perfect measure is as good as any other it is particularly interesting to examine what happens when we change the measure.

<u>3.11 Definitions</u>: Let K be a hyperstonean space. An <u>intrinsic null point</u> in K is a point k such that every neighbourhood of k contains uncountably many disjoint non-empty open sets. If m is a perfect measure on K, an <u>m-null point</u> is a point k such that every neighbourhood of k has infinite m-measure.

Since a non-empty open set has positive m-measure for every perfect measure m, each intrinsic null point is an m-null point. The converse forms part c) of the following proposition which also explains why they are called null points.

3.12 Proposition: Let $X \subset \int_K^p X_k \, dm$ be an integral module.

a) X_k is the null space if and only if k is an m-null point.

b) If Y is an integral module in the direct integral $\int_K^p Y_k \, dm'$ such that $X \cong Y$ as Banach C(K)-modules, then $X_k \cong Y_k$ for all k which are neither m- nor m'-null points.

c) If k is an m-null point for all perfect measures m then k is an intrinsic null point.

Proof: a) If X_k is not the null space there is an element x in X such that $\|x(k)\| = 1$. By continuity there is a neighbourhood D of k such that $\|x(l)\| \geq \frac{1}{2}$ for $l \in D$. By the integrability of $\|x(.)\|^p$ D has finite m-measure. On the other hand suppose D is a (w.l.o.g. clopen) neighbourhood of k with finite m-measure. Let

$A := \{x \mid x \in X, \text{ supp } x \subset D, \chi_{\text{supp } x} \leq \|x(.)\| \leq 2\chi_{\text{supp } x}\}$. Then A has properties (ii) and (iii) in the existence lemma. Since every non-empty clopen set contains a point whose component space is not the null space, A also has property (i). By the existence lemma there is an element $x \in A$ with supp x = D. It follows that $x(k) \neq 0$ and thus that X_k is not the null space.

b) Since every perfect measure is absolutely continuous w.r.t. every other, there are (w.l.o.g. continuous, s. lemma 3.4) numerical functions dm/dm' and dm'/dm such that $m'(G) = \int_G (dm'/dm) . dm$ and vice versa for every Borel set G in K. Let k be a point in K which is neither an m- nor an m'-null point. We assume that dm'/dm is finite at k. Since $dm/dm' = \frac{1}{dm'/dm}$ and the situation is perfectly symmetrical this involves no loss of generality. Because of the continuity of dm'/dm there is a clopen set D containing k on which dm'/dm is finite. Let $f = \chi_D (dm'/dm)^{1/p}$, a C(K)-function. Suppose x is an element of X and y the corresponding element of Y. For every Borel set $G \subset D$ $\int_G \|fy(l)\|^p \, dm' = \int_G \|fx(l)\|^p \, dm$ (because of the isometry as C(K)-modules) $= \int_G \|x(l)\|^p (dm'/dm) \, dm = \int_G \|x(l)\|^p \, dm'$. It follows that $\|fy(\cdot)\| = \|x(\cdot)\|$ on D and in particular that $\|fy(k)\| = \|x(k)\|$.

Define a mapping Φ from X_k into Y_k as follows. If x_k is an element in X_k and x an element of X with $x(k) = x_k$ let $\Phi(x_k) = fy(k)$ whereby y is the element of Y corresponding to x. If two elements of X both take the value x_k at k then their difference takes the value 0. By what we have just shown the difference of f times the corresponding element in Y also takes the value 0 at k. The mapping is thus well-defined, is trivially linear and, by the above norm equality, an isometry. It remains to show that Φ is onto.

Since k is neither an m-null point nor an m'-one, there is a clopen neighbourhood G of k with both m(G) and m'(G) finite. W.l.o.g. $G \subset D$. Consider the net of clopen subsets $G_\alpha \subset G$ on which f is bounded away from zero. Since dm'/dm is almost everywhere non-zero (otherwise there would be an open set with zero m'-measure) $(\bigcup_\alpha G_\alpha)^- = G$. Let y_k be an element of Y_k and y an element of Y with $y(k) = y_k$ and $\|y(.)\|$ everywhere finite. For every α, $\chi_{G_\alpha} \frac{1}{f}$ is a C(K)-function

so that $\chi_{G_\alpha} \frac{1}{f} y$ exists in Y. $N(\chi_{G_\alpha} \frac{1}{f} y)^p = \int_{G_\alpha} (\frac{1}{f} \| y(.) \|)^p \, dm' =$
$= \int_{G_\alpha} \| y(.) \|^p \, dm \leq \int_G \| y(.) \|^p \, dm$ which is finite. It follows that the net $(\chi_{G_\alpha} \cdot \frac{1}{f} \cdot y)$ converges to an element z in Y. Since z is $\frac{1}{f} \cdot y$ on each G_α, fz is equal to y on each G_α and thus everywhere in G. In particular $fz(k) = y_k$. Let x be the element in X corresponding to z. $\Phi(x(k)) = fz(k) = y_k$. Φ is onto.

c) Suppose k is not an intrinsic null point. Then there is a clopen neighbourhood D of k such that every family of pairwise disjoint clopen subsets is at most countable. Let m be some perfect measure on K. Consider all families of disjoint non-empty clopen subsets of D with finite m-measure, ordered by inclusion. These clearly have the chain property so that, by Zorn's Lemma, there is a maximal family , say Q, containing of course at most countably many sets. We number these sets $D_1, D_2, \ldots$ and so on, possibly terminating. Define a function f as follows. $f(t) = 1$ for $t \notin D$, $f(t) = \frac{1}{2^n m(D_n)}$ for $t \in D_n$. Since every clopen set contains a clopen set with finite m-measure, D is the closure of the union of the D_n's. This means that f is defined on an open set which is dense in K and can thus be extended to a continuous function on K. We call this extended function f, too. Define a new measure m' on K by $m'(G) = \int_G f \, dm$ for every Borel set G. Since D is the closure of the union of the D_n's every open set contains a point where f is non-zero. It follows that every open set has positive m'-measure. This is (i) in the definition of a perfect measure, (ii) and (iii) are trivially fulfilled. $m'(D) = \int_D f \, dm =$
$\sum_{n \in \mathbb{N}} \frac{m(D_n)}{2^n m(D_n)} = 1$. k is not an m'-null point. □

We close this section with some purely technical results which we shall need later.

3.13 Lemma: Let $X \subset \int_K^p X_k \, dm$ be an integral module. Let B be a clopen set with finite m-measure, k a point in B and $x_k \in X_k$ an element with norm 1. Then there is an $x \in X$ such that $x(k) = x_k$ and $\|x(.)\| = \chi_B$.

Proof: By lemma 3.10 there is an element $x_o \in X$ such that $x_o(k) = x_k$. Let D be a clopen subset of B such that $\frac{1}{2} \leq \|x_o(l)\| \leq 2$ for $l \in D$. Define the function $f \in C(K)$ by $f(l) = \frac{1}{\|x_o(l)\|}$ for $l \in D$, $f(l) = 0$ for $l \notin D$. Then fx_o is an element of X with $\|fx_o(.)\| = \chi_D$ and $fx_o(k) = x_k$. We apply the existence lemma to the following set

$A = \{x \mid x \in X, \|x(.)\| = \chi_C$ for some $C \subset B \setminus D\}$. Then conditions (ii) and (iii) of the existence lemma are trivially satisfied and (i) follows exactly as the existence of x_o with $\operatorname{supp} x_o \subset B$. It follows that there is an element $x_1 \in X$ with $\|x_1(.)\| = \chi_{B\setminus D}$. Let $x = x_o + x_1$. □

3.14 Corollary: Let $X \subset \int_K^p X_k \, dm$ be an integral module. The set $\mathcal{T} := \{x \mid x \in X, \|x(.)\|$ is a step function$\}$ is dense in X.

Proof: Although there is an elementary proof which does not make use of the existence lemma, we apply lemma 3.13 for convenience.

Let $x \in X$ and $\varepsilon > 0$. Let B be a clopen set with a) $N(\chi_{K\setminus B}x) < \frac{\varepsilon}{2}$, b) $0 < x(k) < \infty$ for $k \in B$, c) $m(B) < \infty$. For every $k \in B$ there is an $x^k \in X$ such that $\|x^k(.)\| = \chi_B$ and $x^k(k) = x(k)$ (by lemma 3.13).

It follows that there is a clopen set $B_k \subset B$ such that $\|x^k(l) - x(l)\| < \frac{\varepsilon}{2 \cdot m(B)^{1/p}}$ for $l \in B_k$. By compactness there are a finite number of (w.l.o.g. pairwise disjoint) B_k's, say $B_{k_1}, \ldots, B_{k_n}$, such that

$B = \bigcup_{i=1}^{n} B_{k_i}$. Let $y = \sum_{i=1}^{n} \chi_{B_{k_i}} x^{k_i} \in X$. Then $\|y(\ )\|$ is a step function and $N(x-y)^p < \frac{\epsilon^p}{2^p} + \int_B \|x(k)-y(k)\|^p dm < \frac{\epsilon^p}{2^p} + \frac{\epsilon^p}{2^p} \leq \epsilon^p$. □

<u>3.15 Corollary</u>: Let $X \subset \int_K^p X_k \, dm$ be an integral module. Then there are elements x_α $(\alpha \in I)$ in X such that $m = \sum_\alpha m_{x_\alpha}$ (i.e. all perfect measures can be constructed as in section A).

<u>Proof</u>: Let $(B_\alpha)_{\alpha \in I}$ be a maximal family of disjoint clopen sets with finite m-measure (chosen by Zorn's Lemma). Then $K = (\bigcup_\alpha B_\alpha)^-$. By lemma 3.13 there are elements x_α in X such that $\|x_\alpha(.)\| = \chi_{B_\alpha}$ for each α. Since $m_x = \|x(.)\|^p m$ for all x, $m_{x_\alpha} = \chi_{B_\alpha} m$ for each α. Define a measure m' by $m'(G) = \sum_\alpha m_{x_\alpha}(G) = \sum_\alpha m(G \cap B_\alpha)$ for each Borel set G. Since both m and m' vanish on sets of first category it suffices to show that they agree on clopen sets. Let B be a clopen set with finite m-measure. Then $\sum_\alpha m(B \cap B_\alpha)$ has at most countably many non-zero terms, say those involving $\alpha_1, \ldots, \alpha_n, \ldots$, and therefore $m'(B) = \sum_\alpha m(B \cap B_\alpha) = \sum_{i=1}^{\infty} m(B \cap B_{\alpha_i}) = m(B \cap \bigcup_{i=1}^{\infty} B_{\alpha_i}) = m(B \cap (\bigcup_{i=1}^{\infty} B_{\alpha_i})^-)$. But $B \setminus (\bigcup_{i=1}^{\infty} B_{\alpha_i})^-$ is a clopen set and is then either empty or intersects a B_α. But then $B \cap B_\alpha \neq \emptyset$ so that there would be another non-zero term in the sum. It follows that this set is empty and thus that $B \cap (\bigcup_{i=1}^{\infty} B_{\alpha_i})^- = B$. $m'(B) = m(B)$ and since each clopen set contains a clopen set with finite m-measure m' and m agree on all clopen sets. $m' = m$. □

For a discussion of a measure-theoretical approach to integral modules, the reader is referred to appendix 3.

Chapter 4: The classical L^p-spaces

In this chapter we shall apply the theory of the foregoing chapters to obtain some characterizations of the classical L^p-spaces. Since the L^p-projections were defined in such a way as to satisfy a norm identity similar to the definition of the L^p-norm it is to be expected that there will be many L^p-projections on L^p-spaces and that L^p-spaces can be characterized as those having, in some sense, a maximal number of L^p-projections.

Before proceeding to the classical spaces we prove some results concerning arbitrary Banach spaces which we shall need for the characterization and which are also of some interest in their own right.

4.1 Definition: Let X be a Banach space and M a subset of $[X]$ whose elements commute with one another. An M-cycle is a closed subspace of X which is invariant for every element of M. An M-ideal is a closed subspace of X which is invariant for every element of $(M)_{COMM}$.

Since we have required that the elements of M commute with one another, every M-ideal is an M-cycle. If M consists only of Id and 0 the M-cycles are the closed subspaces and the only M-ideals are X and $\{0\}$. If $(J_\alpha)_{\alpha\in I}$ is a family of M-cycles (M-ideals) then $\bigcap_{\alpha\in I} J_\alpha$ and $(\text{lin} \bigcup_{\alpha\in I} J_\alpha)^-$ are also M-cycles (M-ideals). In particular we can define the M-cycle generated by an element x as the smallest M-cycle containing x. We write this cycle $S(M;x)$.

We are naturally interested in the case where M is a complete Boolean algebra of L^p-projections, $1 \leq p < \infty$. In this case the cycles generated by single elements have a particularly simple form as shown in the following lemma.

<u>4.2 Lemma</u>: Let X be a Banach space, $1 \leq p < \infty$, and $\mathfrak{A}$ a complete Boolean algebra of L^p-projections on X. Then

$$S(\mathfrak{A};x) \cong L^p(m_x)$$

for all $x \in X$, whereby m_x is the Borel measure on the Stonean space of $\mathfrak{A}$ defined by x as in 3.1.

<u>Proof</u>: As in chapter 3 let C be the closed subspace of [X] generated by $\mathfrak{A}$, m a perfect measure on the Stonean space Ω of $\mathfrak{A}$ and $[\cdot]$ the norm resolution for X taking values in $C^p(\Omega;m)$. Since $S(\mathfrak{A};x)$ contains Ex for all E in $\mathfrak{A}$ it follows that $Tx \in S(\mathfrak{A};x)$ for all T in C. However, for E in $\mathfrak{A}$ and T in C $E(Tx) = (ET)x$ so that $\{Tx \mid T \in C\}$ is a linear space which is invariant for every element of $\mathfrak{A}$. It follows that $S(\mathfrak{A};x) = \{Tx \mid T \in C\}^-$. Since m_x is finite $\mathcal{L}^\infty(m_x) \subset \mathcal{L}^p(m_x)$ and C is identified with $C(\Omega) \subset \mathcal{L}^\infty(m_x)$. We can thus define a mapping φ from $S(\mathfrak{A};x)$ into $L^p(m_x)$ as follows: Let $\varphi(Tx) :=$ (the equivalence class of T in $L^p(m_x)$) for T in C. This is well-defined since $Tx = Sx \Rightarrow (T-S)x = 0 \Rightarrow |T-S|[x] = 0 \Rightarrow |T-S| = 0$ m_x-almost everywhere. φ is clearly linear and, since $\|T\|_p^p = \int_\Omega |T|^p dm_x = \int_\Omega |T|^p [x]^p dm = \int_\Omega [Tx]^p dm = \|[Tx]\|_p^p = \|Tx\|^p$, φ is an isometry. We can thus extend φ to an isometry from $S(\mathfrak{A};x)$ into $L^p(m_x)$. Since the continuous functions are dense in $L^p(m_x)$, φ is onto. □

<u>4.3 Corollary</u>: Let K be an extremally disconnected compact topological space, $(X_k)_{k \in K}$ a family of Banach spaces and X an integral module in the direct integral of the X_k with respect to m, where m

is a perfect measure on K. Then there is a sub-module Y of X which is isometrically isomorphic to $C^p(K;m)$ in such a way that the restriction of the isometry to the positive cone of $C^p(K;m)$ is the mapping which maps an element y into its associated scalar function $\tilde{y}$.

<u>Proof</u>: By Cor. 3.15 m is $\sum_I m_{x_\alpha}$ for some family $(x_\alpha)_{\alpha \in I}$ in X, and the measures have pairwise disjoint support in K. Let Y := $(\mathrm{lin} \bigcup_I S(\mathfrak{A};x_\alpha))^-$ whereby $\mathfrak{A}$ is the Boolean algebra of characteristic projections. Since each $S(\mathfrak{A};x_\alpha)$ is isometric to $L^p(m_{x_\alpha})$ and the measures have disjoint support, $Y \cong \sum_I{}^p L^p(m_{x_\alpha}) \cong L^p(m) \cong C^p(K;m)$. That the isometry has the required properties is clear from the construction used in lemma 4.2. □

In an arbitrary Banach space any one-dimensional subspace is the range of a projection with norm 1. The cycles $S(\mathfrak{A};x)$ with which we dealt in the foregoing lemma may be considered as being in some sense "one-dimensional" cycles. Cohen and Sullivan [CS] have shown that in a smooth reflexive space cycles of the form $S(\mathfrak{A};x)$ where $\mathfrak{A}$ is a complete Boolean algebra of L^p-projections, are the range of a norm 1 projection which commutes with each element of $\mathfrak{A}$. The following theorem shows that this is true in all Banach spaces.

<u>4.4 Theorem</u>: Let $\mathfrak{A}$ be a complete Boolean algebra of L^p-projections on a Banach space X, $1 \leq p < \infty$. For every x in X there is a norm 1 projection F in $(\mathfrak{A})_{COMM}$ such that $FX = S(\mathfrak{A};x)$.

<u>Proof</u>: If x = 0 we can put F = 0 also. So w.l.o.g. $\|x\| = 1$.

As in the case of an integral module representation, let Ω be the Stonean space of $\mathfrak{A}$, m a perfect measure on Ω and $[\cdot]$ the norm resolution for X taking values in $C^p(\Omega;m)$. Since $S(\mathfrak{A};x)$ is canonically isomorphic to $L^p(m_x)$, $\int : f \mapsto \int f\, dm_x$ defines a linear functional with norm 1 on $S(\mathfrak{A};x)$. Let $D := \operatorname{supp} m_x \subset \Omega$. $E_D X$ is an $\mathfrak{A}$-cycle and thus $E_D X \supset S(\mathfrak{A};x)$. $\int \circ E_D$ is thus a linear functional with norm 1 on $S(\mathfrak{A};x) \oplus (Id-E_D)X$.

Let x' be an element of X' with norm 1 which extends $\int \circ E_D$. For each $y \in X$, $f \mapsto x'(fy)$ is a continuous linear functional on $C(\Omega)$. There is therefore a regular Borel measure m^y such that $x'(fy) = \int_\Omega f\, dm^y$ for all $f \in C(\Omega)$. Since x' is an extension of $\int \circ E_D$ m^y vanishes on $\Omega \setminus D$ and, since order convergence of a net (f_α) implies the convergence of $f_\alpha y$ and x' is continuous, m^y is a normal measure and thus absolutely continuous with respect to m_x. There is thus a (w.l.og. continuous) function f^y such that $m^y = f^y m_x$. Suppose there is a clopen set B ($\neq \emptyset$) such that $f^y[x] > [y]$ on B. Since f^y is clearly 0 where $[y]$ is, we can define a function f by $f := [x]/[y]$ on B and 0 elsewhere.

$\|\chi_{\Omega\setminus B}x + fy\|^p = \int_{\Omega\setminus B}[x]^p dm + \int_B f^p[y]^p dm = \|x\|^p = 1$. But on the other hand $x'(\chi_{\Omega\setminus B}x + fy) = x'(\chi_{\Omega\setminus B}x) + x'(fy) = \int_{\Omega\setminus B} dm_x + \int_B f f^y dm_x > \int_\Omega dm_x = 1$. This contradicts the fact that x' has norm 1. Thus $f^y[x] \leq [y]$ everywhere (both sides are continuous). Analogously $f^y[x] \geq -[y]$. It follows that $\int_\Omega |f^y|^p dm_x = \int_\Omega |f^y[x]|^p dm \leq \int_\Omega [y]^p dm = \|y\|^p$. f^y is thus in $L^p(m_x)$ ($\cong S(\mathfrak{A};x)$) and $\|f^y\|_p \leq \|y\|$.

We define a mapping F from X into $S(\mathfrak{A};x)$ by means of $y \mapsto f^y$.

Since x' is linear and has norm 1, F is also linear and $\|F\| \leq 1$. If $g \in C(\Omega)$ then $\int_\Omega h\, f^{gx}\, dm_x = x'(hgx)$ for all h in $C(\Omega)$. On the other hand $x'(hgx) = \int_\Omega hg\, dm_x$. It follows that $f^{gx} = g$ and thus that $Fy = y$ for all y in $S(\mathfrak{A};x)$ of the form $f = gx$ for some g in $C(\Omega)$. However elements of this form are dense in $S(\mathfrak{A};x)$ so that, by continuity, $Fy = y$ for all y in $S(\mathfrak{A};x)$. F is thus a projection with norm 1 from X onto $S(\mathfrak{A};x)$. Since for $g \in C(\Omega)$, $y \in X$ on the one hand $\int_\Omega hgf^y dm_x = x'(hgy)$ for all $h \in C(\Omega)$ and on the other hand $\int_\Omega hf^{gy} dm_x = x'(hgy)$ for all $h \in C(\Omega)$, $gf^y = f^{gy}$ and F commutes with $(\mathrm{lin}\ \mathfrak{A})^-$, in particular with the projections in $\mathfrak{A}$. □

This result is the key to our characterization of L^p-spaces by the maximality of their L^p-structure, for if the L^p-projections form a maximal Boolean algebra then the projections constructed in the theorem must also be L^p-projections since the maximality of the algebra implies that there is no other projection which commutes with each L^p-projection.

We turn our attention now to ideals. As with cycles we are interested in the case where the relevant subset of [X] consists of L^p-projections. Since we have already used the term "L^p-summand" to refer to a subspace which is the range of an L^p-projection it is to be expected that the two concepts will be related in some close way. This is the subject of the following proposition.

<u>4.5 Proposition</u>: Let $\mathfrak{A}$ be a complete Boolean algebra of L^p-projections on a Banach space X, $1 \leq p < \infty$. A subspace of X is an $\mathfrak{A}$-

ideal if and only if it is the range of a projection in $\mathfrak{A}$.

<u>Proof</u>: If $J = EX$ with E in $\mathfrak{A}$ then $T \in (\mathfrak{A})_{COMM}$ implies that T commutes with E and thus that $TJ = TEX = ETX \subset EX = J$. J is thus an $\mathfrak{A}$-ideal.

Conversely, suppose that J is an $\mathfrak{A}$-ideal in X. As we saw in the last chapter we can represent X as an integral module over (Ω, m) where Ω is the Stonean space of $\mathfrak{A}$ and m is an arbitrary perfect measure. In particular, there is a norm resolution for X taking values in $C^p(\Omega;m)$. We write this norm resolution $[\cdot]$. The proof is divided into three steps:

(a) If $x \in J$, $y \in X$, $a \in \mathbb{R}$ and $[y] \leq a[x]$ then y is also in J.

We define a map φ from $S(\mathfrak{A};x)$ into $S(\mathfrak{A};y)$ by $\varphi(Tx) = Ty$ for T in C ($:= (\text{lin}\,\mathfrak{A})^-$). φ is welldefined and continuous since $[y] \leq a[x]$ and can thus be extended by continuity to all of $S(\mathfrak{A};x)$. If F is the projection whose existence was proved in the last theorem which maps X onto $S(\mathfrak{A};x)$ then $\varphi F \in (\mathfrak{A})_{COMM}$ and $\varphi F(x) = y$. Since J is an $\mathfrak{A}$-ideal, y is in J.

(b) If $x \in J$, $y \in X$ and $\text{supp}\,[y] \subset \text{supp}\,[x]$ then $y \in J$.

For all $n \in \mathbb{N}$ let $B_n := \{k \mid [y](k) < n[x](k)\}^-$. Since $[x]$ and $[y]$ are continuous, B_n is the closure of an open set and thus clopen. Let E_n be the projection associated with the set B_n. Since $(\bigcup_{n=1}^{\infty} B_n)^- = \text{supp}[x] \supset \text{supp}\,[y]$ we have that $Ey = y$, where E is the projection associated with $(\bigcup_{n=1} B_n)^-$. But E is the supremum of the E_n's and thus also limit in the strong operator topology. It follows that $\lim E_n y = Ey = y$. But since $[E_n y] \leq$

$n[x]$, $E_n y$ is in J by (a). Since J is closed, y lies also in J.

(c) $J = EX$ for some $E \in \mathfrak{U}$

Let D be the set $(\bigcup_{x \in J} \operatorname{supp}[x])^-$. D is the closure of an open set and thus clopen. Let E be the corresponding projection in $\mathfrak{U}$. Since for all x in J $\operatorname{supp}[x] \subset D$ we have that $Ex = x$ for all x in J and thus that $EX \supset J$. But suppose that $y \in EX$. For every $x \in J$ $\operatorname{supp}[E_{\operatorname{supp}[x]} y] \subset \operatorname{supp}[x]$ where $E_{\operatorname{supp}[x]}$ is the projection in $\mathfrak{U}$ corresponding to the clopen set $\operatorname{supp}[x]$. It follows from (b) that $E_{\operatorname{supp}[x]} y$ is in J for every x in J. However, as D is $(\bigcup_{x \in J} \operatorname{supp}[x])^-$, E is the supremum of the $E_{\operatorname{supp}[x]}$'s and thus their limit in the strong operator topology. Since J is closed, Ey is in J. As $y \in EX$, $Ey = y$ and thus $EX = J$. □

We now come to our characterization of L^p-spaces as those Banach spaces whose L^p-structure is maximal. At first glance there seems to be more than one way of interpreting the word maximal. Firstly the L^p-projections form a Boolean algebra of projections in $[X]$, so that one can interpret the word maximal to mean that the Boolean algebra of L^p-projections is a maximal Boolean algebra of projections in $[X]$. On the other hand the closed linear hull of the set of L^p-projections is a commutative sub-algebra of $[X]$ so that one can also interpret the word maximal to mean that this algebra is a maximal commutative sub-algebra of $[X]$. This is apparently a stronger maximality than that as projection algebra since it implies that there is no operator commuting with the L^p-projections and not just that there is no other projection. However, as the theorem shows the two concepts are identical and the proof depends naturally on theorem 4.4 which showed the existence of sufficiently many projections.

<u>4.6 Theorem</u>: Let $\mathfrak{A}$ be a complete Boolean algebra of L^p-projections on a Banach space X, $1 \le p < \infty$. The following are equivalent:

(i) $(\mathfrak{A})_{COMM} = (\text{lin } \mathfrak{A})^-$

(ii) $\mathfrak{A}$ is maximal as Boolean algebra of bounded projections

(iii) $S(\mathfrak{A};,x)$ is an $\mathfrak{A}$-ideal for all x in X

(iv) Every $\mathfrak{A}$-cycle is an $\mathfrak{A}$-ideal

(v) $X \cong L^p(m)$ for a perfect measure on the Stonean space of $\mathfrak{A}$

<u>Proof</u>:

(i) ⇒ (ii): obvious (cf. 2.2)

(ii) ⇒ (iii): By theorem 4.4 $S(\mathfrak{A};x)$ is the range of a projection F with norm ≤ 1 which commutes with all the projections in $\mathfrak{A}$. Since $\mathfrak{A}$ is maximal F must lie in $\mathfrak{A}$. $S(\mathfrak{A};x) = FX$ is then an $\mathfrak{A}$-ideal.

(iii) ⇒ (iv): Let J be an $\mathfrak{A}$-cycle. Then $S(\mathfrak{A};x) \subset J$ for all $x \in J$. It follows that $J = (\text{lin} \bigcup_{x \in J} S(\mathfrak{A};x))^-$. Since the individual $S(\mathfrak{A};x)$'s are $\mathfrak{A}$-ideals it follows that J is also an $\mathfrak{A}$-ideal.

(iv) ⇒ (v): Let m be a perfect measure on the Stonean space Ω of $\mathfrak{A}$. We shall show that $X \cong L^p(m)$. Suppose that $m = \sum_I m_{x_\alpha}$ for some disjoint x_α in X, $\alpha \in I$. For simplicity's sake we write m_α for m_{x_α}, D_α for supp m_α and E_α for the projection in $\mathfrak{A}$ associated with D_α. Since $m_\alpha \le m$ for all α we have the natural mapping $L^p(m) \to L^p(m_\alpha)$

and the spaces $S(\mathfrak{A};x_\alpha)$ and $L^p(m_\alpha)$ can be identified as in lemma 4.2. For $f \in L^p(m)$ we have $\int_\Omega |f|^p dm = \sum_\alpha \int_{D_\alpha} |f|^p dm_\alpha = \sum_\alpha \|f_\alpha\|^p$. It follows that for $f \in L^p(m)$ the sum $\sum_\alpha f_\alpha$ exists in X. We define a mapping φ from $L^p(m)$ into X by $\varphi(f) := \sum_\alpha f_\alpha$:

a) φ is linear

b) Since the D_α are disjoint and $(\bigcup_\alpha D_\alpha)^- = \Omega$ the norm of $\sum_\alpha f_\alpha$ in X is $(\sum_\alpha \|f_\alpha\|^p)^{1/p}$. It follows that φ is an isometry.

c) The range of φ (= $(\mathrm{lin} \bigcup_\alpha S(\mathfrak{A};x_\alpha))^-$) is an $\mathfrak{A}$-cycle and thus an $\mathfrak{A}$-ideal. By proposition 4.5 the range of φ is EX for some E in $\mathfrak{A}$. Since however x_α is in the range of φ for each α and the union of the D_α's is dense in Ω the clopen set associated with E must be Ω itself. E is thus Id and φ is onto.

(v) $\Rightarrow$ (i): Let $X \cong L^p(m) \cong C^p(\Omega;m)$ and T an element in $(\mathfrak{A})_{\mathrm{COMM}}$.

a) $f \in C^p(\Omega;m)$. Suppose that there is a point k with $|Tf(k)| > \|T\||f(k)|$. Then there is a nonvoid clopen set D such that $|Tf(l)| > \|T\||f(l)|$ for all l in D. But then $\int |T(\chi_D f)|^p dm = \int |\chi_D Tf|^p dm > \|T\|^p \int_D |f|^p dm$ in contradiction to $\|T(\chi_D f)\| \leq \|T\| \|\chi_D f\|$. It follows that for all $f \in C^p(\Omega;m)$ $|Tf| \leq \|T\||f|$.

b) Let f and g be two $C^p(\Omega;m)$-functions that are finite and non-zero at the point k. Then $(g(k)f - f(k)g)(k) = 0$. By a) it follows that $T(g(k)f - f(k)g)(k) = 0$. But since T is linear this means that $g(k)Tf(k) - f(k)Tg(k) = 0$. Or in other words $Tf(k)/f(k) = Tg(k)/g(k)$.

c) Let B be the set of m-null points in Ω. Then there is a $C^p(\Omega;m)$-function which is finite and non-zero at the point k if and only if k is not in B. We define a function T^* from $\Omega \setminus B$ into $\mathbb{R}$ by $T^*(k) := Tf(k)/f(k)$ where f is a $C^p(\Omega;m)$-function which is finite and non-zero at k. By a) and b), T^* is well-defined and bounded. T^* is continuous, because we can use the same function f in a neighbourhood of k. Since $\Omega \setminus B$ is a dense open set it follows that we can extend T^* to a $C(\Omega)$-function. By b) T then has the action of almost everywhere multiplication with this $C(\Omega)$-function and these are exactly the operators in $(\mathrm{lin}\ \mathfrak{A})^-$. □

To close ths chapter we show the concrete form of the L^p-projections on an arbitrary L^p-space, $L^p(\mu)$. The following results are equivalent to results of Sullivan [S2] which, however, are formulated slightly differently.

First of all, it is clear that in any L^p-space, $L^p(\mu)$, the mapping $f \to f\chi_B$ (B a μ-measurable set) is an L^p-projection. We call such projections <u>characteristic</u>. In general the characteristic projections, although they clearly form a σ-complete Boolean algebra, do not form a complete one. The reason for this is that the measurable sets are only closed under countable operations. However, they will be contained in a complete Boolean algebra for which then the following proposition holds.

<u>4.7 Proposition</u>: If X is an L^p-space (i. e. $X = L^p(\mu)$ for some measure μ) and $\mathfrak{A}$ a complete Boolean algebra of L^p-projections con-

taining the characteristic projections, then the five equivalent conditions of theorem 4.6 hold ($1 \leq p < \infty$).

<u>Proof</u>: It is of course sufficient to show that one of them holds and we choose (iii).

Let f be in X and E be the characteristic projection $g \to \chi_{\text{supp } f}\, g$. Then $S(\mathfrak{A};f) \subset EX$. Let g be any element of EX. For every $n \in \mathbb{N}$ let B_n be the set $\{t \mid (1/n)|g(t)| < |f(t)|\}$ and $h_n := \chi_{B_n}(g/f)$ (i. e. $h_n(t) := g(t)/f(t)$ for t in B_n and $h_n(t) = 0$ for $t \notin B_n$). Then $h_n \in L^\infty(\mu)$ for each n and $h_n f \chi_{B_n} = g \chi_{B_n}$. But supp $f \supset$ supp g and therefore $\bigcup_{n=1}^{\infty} B_n =$ supp g μ-a.e. . We thus have $\lim h_n f = g$. Since $\mathfrak{A}$ contains the characteristic projections, C ($= (\text{lin } \mathfrak{A})^-$) contains the operators $h \to hh_n$ ($h \in X$) for all n. Thus $h_n f \in S(\mathfrak{A};f)$ for all n and, since $S(\mathfrak{A};f)$ is closed, $g \in S(\mathfrak{A};f)$. $S(\mathfrak{A};f)$ is therefore equal to EX, an $\mathfrak{A}$-ideal. □

Theorem 4.6 is thus a characterization of arbitrary L^p-spaces and not merely $L^p(m)$'s where m is a perfect measure (though it naturally shows that an arbitrary L^p-space can in fact be put in this form).

It follows now from (i) of 4.6 that there is only one complete Boolean algebra of L^p-projections containing the characteristic ones. Since the strong closure of a Boolean algebra of L^p-projections is itself a complete Boolean algebra of such projections, every L^p-projection on an L^p-space which commutes with the characteristic projections (by 1.3 this condition is satisfied for every L^p-projection if $p \neq 2$) is the strong limit of characteristic projections. These projections can also be described in the following way:

<u>4.8 Definition</u>: Let $S(\mu)$ be the set of all σ-finite subsets (w.r.t. μ) of a measure space μ. A <u>pseudocharacteristic function</u> (PCF) is a function $P: S(\mu) \to S(\mu)$ such that $P(B \cap D) = P(B) \cap D$ μ-a.e. for all B, D in $S(\mu)$. Each PCF defines a <u>pseudocharacteristic projection</u> (PCP) $E_P : L^p(\mu) \to L^p(\mu)$ by virtue of $E_P(f) := \chi_{P(\text{supp } f)} f$. Clearly, a PCP is a projection which commutes with the characteristic projections. Since $E_P f$ and $f - E_P f$ have disjoint supports, a PCP is an L^p-projection.

<u>4.9 Proposition</u>: Let X be an L^p-space, $X = L^p(\mu)$ $(1 \leq p < \infty)$, E an L^p-projection on X which commutes with the characteristic projections. Then E is pseudocharacteristic. In particular, for $p \neq 2$ every L^p-projection is pseudocharacteristic

<u>Proof</u>: We define $P : S(\mu) \to S(\mu)$ by $P(\text{supp } f) := \text{supp } Ef$ for f in $L^p(\mu)$.

a) <u>P is well-defined</u>

Since a measurable set is σ-finite if and only if it is the support of an $L^p(\mu)$-function P is defined on all of $S(\mu)$. On the other hand if supp f = supp g then, as in the proof of 4.7, $f \in S(\mathfrak{A};g)$, $g \in S(\mathfrak{A};f)$ (where $\mathfrak{A}$ is the complete Boolean algebra generated by the characteristic projections), so that $Ef \in S(\mathfrak{A};Eg)$, $Eg \in S(\mathfrak{A};Ef)$ and supp Ef = supp Eg μ-a.e. .

b) <u>P is a PCF</u>

$P(B \cap C) = P(\text{supp } \chi_C f) = \text{supp } E(\chi_C f) = \text{supp } \chi_C Ef = C \cap \text{supp } Ef$
$= C \cap P(B)$ μ-a.e. (whereby supp $f = B$)

c) <u>$E = E_P$</u>

Let $f \in L^p(\mu)$ and $B = P(\text{supp } f)$. Then $\chi_B Ef = Ef = E(\chi_B f) = EE_P f$. $E_P f \in S(\mathfrak{A}; Ef)$ implies that there is a sequence $(H_n)_{n\in\mathbb{N}}$ in lin $\mathfrak{A}$ such that $H_n Ef \to E_P f$. Then $EH_n Ef \to EE_P f = Ef$. But

$EH_n Ef = H_n Ef \to E_P f$. □

We thus have an explicit description of all L^p-projections on an $L^p(\mu)$ $(1 \leq p < \infty, p \neq 2)$. In $L^2(\mu)$ every orthogonal projection is an L^2-projection, though in this case the pseudocharacteristic projections do form a maximal Boolean algebra of L^2-projections as a suitable modification of the above results easily shows.

We can also give an explicit description of the Cunningham p-algebra $(p \neq 2)$:

<u>4.10 Corollary</u>: The Cunningham p-algebra of an L^p-space, $L^p(\mu)$ $(p \neq 2, p < \infty)$, is isometric to $(L^1(\mu))'$ $(\cong L^\infty(\mu)$ if μ is localizable).

<u>Proof</u>: If we represent $L^p(\mu)$ as an integral module with respect to the algebra of pseudo-characteristic projections, the conditions of theorem 4.6 are satisfied. The Cunningham p-algebra is identified with $C(\Omega) \cong L^\infty(m) \cong (L^1(m))'$ and $L^p(\mu) \cong L^p(m)$ (condition (v)). It follows that $L^1(\mu) \cong L^1(m)$ and thus that the Cunningham p-algebra is isometric to $(L^1(\mu))'$. □

Chapter 5: Integral Modules and Duality

In this chapter we shall compare integral module representations of a Banach space X and its dual. Let $1 \leq p < \infty$ and $\frac{1}{p} + \frac{1}{p'} = 1$.

If μ is a σ-finite, positive, not necessarily atomic measure, we know that

$$(L^p(\mu))' \cong L^{p'}(\mu)$$

by virtue of the scalar product

$$\langle \varphi, f \rangle := \int \varphi f \, d\mu .$$

For vector valued spaces, in the atomic case $X = \prod^p_{i \in I} X_i$, one can easily show the analogous result

$$(\prod^p_{i \in I} X_i)' \cong \prod^{p'}_{i \in I} X_i'$$

by virtue of the scalar product

$$\langle \varphi, x \rangle := \sum_{i \in I} \varphi(i)(x(i)) .$$

Thus the following question arises. Let X be an arbitrary p-integral module with components X_k ($k \in K$). Does the dual X' of the Banach space X have a p'-integral module representation on the same measure space (K,m) (resp. a function module representation on the same K) with the components X_k', by virtue of a scalar product

$$\langle \varphi, x \rangle := \int_K \varphi(k)(x(k)) \, dm \quad ?$$

We shall give an affirmative answer in the case $1 < p < \infty$, with the surprising restriction that in general the components Z_k of the representation of X' are weak-*-dense, norm closed subspaces of X_k'. Only in some special cases can we show that this subspace is all of X_k'. This result is surprising since, in the case of the classical function spaces $L^p(\mu)$, the embedding of $L^p(\mu)'$ into $L^{p'}(\mu)$

requires the relatively deep Radon-Nikodym theorem (which will come into our proof, too), whereas surjectivity is rather easy. For the case p = 1, an analogue result is shown in [Gr] by use of similar methods.

A: The Isometric Embedding $X' \longrightarrow \int_K^{p'} X_k' \, dm$

In the following, let $1 < p < \infty$.

We start with a p-integral module X in a p-direct integral $\int_K^p X_k \, dm$. First we look for a complete Boolean algebra of $L^{p'}$-projections, whose Stonean space is K.

5.1 Proposition: A subset $\mathfrak{A}$ of [X] is a complete Boolean algebra of L^p-projections if and only if $(\mathfrak{A})^t := \{E' \mid E \in \mathfrak{A}\}$ is a complete Boolean algebra of $L^{p'}$-projections in [X'].

Proof: As $E = E''|_X$ for all $E \in [X]$ and $p'' = p$, it suffices to show "$\Longrightarrow$". It is well known that $([X])^t$ coincides with the algebra of weak-*-continuous operators on X'. In 1.6 we have defined $P[X] := \{P \mid P \in [X], P^2 = P, \|P\| \leq 1\}$. As $(.)' : T \longmapsto T'$ is an isometric algebra isomorphism, $(P[X])^t$ coincides with the set $P_\sigma[X']$ of weak-*-continuous projections in [X'] with norm ≤ 1. As $(.)'$ is multiplicative, $(.)'|_{P[X]}$ is monotone, hence an order isomorphism of P[X] onto $P_\sigma[X']$. Therefore $(\mathfrak{A})^t$ is a complete Boolean algebra if $\mathfrak{A}$ is. Let $\mathfrak{A}$ be a complete Boolean algebra of L^p-projections in [X], (E_α') an increasing net in $(\mathfrak{A})^t$. From the remark after 1.6 we have that the strong limit F of (E_α') exists, is an $L^{p'}$-projection and equals the supremum of (E_α') in P[X']. We have to show that

$F \in (\mathfrak{A})^t$. $L^{p'}$-projections in a dual space are weak-*-continuous (2.9), hence F is the supremum of (E_α') in $P_\sigma[X']$. As $(.)'|_{P[X]}$ is an order isomorphism, $F = E'$, where E is the supremum of (E_α') in $P[X]$. $\mathfrak{A}$ is complete in the sense of projection algebras and therefore $E \in \mathfrak{A}$; hence $F \in (\mathfrak{A})^t$. □

$\mathfrak{A}$ and $(\mathfrak{A})^t$ are isomorphic by means of $(.)'$. Therefore K is the Stonean space of $(\mathfrak{A})^t$ and, if for better distinction we denote the image of $f \in C(K)$ in $C_{p'}(X')$ (see 2.1) by S_f instead of T_f, we have

① $$S_f = (T_f)' \quad .$$

Further we know from chapter 3 that X' has a representation as a p'-integral module in a p'-direct integral $\int_K^{p'} Y_k \, dm$. We shall replace the components Y_k by subspaces Z_k of X_k'.

For reasons of symmetry we look at a somewhat more general situation.

5.2 Definition: Let X, Y be Banach spaces. We call a bilinear form $\langle \cdot,\cdot \rangle : X \times Y \longrightarrow \mathbb{R}$ faithful, if the partial mappings at $x \in X$ and $y \in Y$ are bounded, with norm $\|x\|$ and $\|y\|$ resp.. If, in addition, X is a p-integral module and Y is a p'-integral module on the same measure space (K,m), and $\langle fx,y \rangle = \langle x,fy \rangle$ for all $x \in X$, $y \in Y$, $f \in C(K)$, we call the integral modules X and Y dual (by virtue of $\langle \cdot,\cdot \rangle$).

Faithful bilinear forms induce isometric linear mappings from Y into X' and from X into Y'. Note that the ranges of these induced mappings are weak-*-dense in X' and Y' resp., as a subspace Z of a Banach dual is weak-*-dense if and only if the restricted evaluation $x \longrightarrow ev_x|_Z$ from X into Z' is injective (the kernel of this mapping coincides with the polar of Z).

In what follows, X and Y are dual integral modules of $\int_K^p X_k \, dm$ and $\int_K^{p'} Y_k \, dm$ resp. by virtue of a bilinear form $\langle \cdot , \cdot \rangle$.

With the aid of the next lemma we shall provide faithful bilinear forms $\langle \cdot,\cdot \rangle_k : X_k \times Y_k \longrightarrow \mathbb{R}$ which will allow us to interpret Y_k as subspaces of X_k' $(k \in K)$.

5.3 Lemma: Let X, Y, K, $\langle \cdot , \cdot \rangle$ be as above, $x \in X$ and $y \in Y$. The mapping $f \longmapsto \langle fx,y \rangle$ defines a Radon measure on C(K). The induced Borel measure $\mu_{x,y}$ on K is absolutely continuous w.r.t. m; consequently it has a (w.l.o.g. continuous) derivative $d_{x,y}$ w.r.t. m.

Proof: If we define $\nu(f) := \langle fx,y \rangle$, the linear form ν is bounded since $|\langle fx,y \rangle| \le \|f\| \cdot \|x\| \cdot \|y\|$ ($\langle \cdot,\cdot \rangle$ is faithful), and therefore a Radon measure.

Let $\mu := \mu_{x,y}$. The m-null sets are exactly the measurable sets of first category, and each $|\mu|$-null set is a μ-null set. Therefore we have only to show that $|\mu|$ vanishes on nowhere dense sets.

We observe that $|g| \le |f|$ implies $\|gx\| \le \|fx\|$, because $\|gx\|^p = \int_K [gx]^p \, dm = \int_K |g|^p [x]^p \, dm$. Therefore for clopen $B \subset K$ we have

$$|\mu|(B) = |\nu|(\chi_B) = \sup \nu([-\chi_B, \chi_B]) = \sup \{ \langle gx,y \rangle \mid |g| \le \chi_B \} \le$$
$$\le \|y\| \cdot \sup \{ \|gx\| \mid |g| \le \chi_B \} = \|y\| \cdot \|E_B x\| .$$

As $|\mu|$ is regular, for nowhere dense sets $D \subset K$ we get $|\mu|(D) =$
$= \inf \{ |\mu|(B) \mid D \subset B, \text{ B open} \} \le \inf \{ |\mu|(B) \mid D \subset B, \text{ B clopen} \} \le$
$\le \|y\| \cdot \inf \{ \|E_B x\| \mid D \subset B, \text{ B clopen} \} = 0$. The last equation holds, because the infimum $E_\emptyset$ of the decreasing net $(E_B)_{B \supset D, \text{ B clopen}}$ is the strong limit of this net, too (1.6). Because of lemma 3.4 the measure μ, which is absolutely continuous w.r.t. m, has exactly one

continuous derivative. □

Now we can prove the theorem on the representation of the dual space.

5.4 Theorem: Let X, Y, K, $\langle \cdot , \cdot \rangle$ be as above. Then

(i) for all $k \in K$ $\langle x(k), y(k) \rangle_k := d_{x,y}(k)$ defines a faithful bilinear form on $X_k \times Y_k$

(ii) $\langle x,y \rangle = \int_K \langle x(k), y(k) \rangle_k \, dm$

(iii) Y has a representation $\sim$ as p'-integral module in $\int_K^{p'} Z_k \, dm$ with weak-*-dense, norm closed subspaces Z_k of X_k', so that

$$\langle x,y \rangle = \int_K \tilde{y}(k)(x(k)) \, dm$$

where the integrands in (ii) and (iii) are m-almost everywhere defined and continuously extendable.

Note:

Because of the equation (1) a p-integral module X and its dual are dual as integral modules. Therefore 5.4 provides a representation of the dual of X and, if X is a dual space itself, representations of the preduals of X.

Proof: If we have (i), (ii) is trivial from the definition of $d_{x,y}$, and we get (iii) by means of the isometric isomorphisms from Y_k onto Z_k, induced by the faithful bilinear forms $\langle \cdot , \cdot \rangle_k$.

We shall prove (i) in two steps. Let $k \in K$. First of all, we may assume that k is not an m-null point; otherwise $d_{x,y}(k) = 0$, because $d_{x,y} \in C^1(K;m)$, and $X_k = \{0\} = Y_k$.

I. For all $(x,y) \in X \times Y$ with $(x(k), y(k)) \in X_k \times Y_k$

$|d_{x,y}(k)| \leq [x](k) \cdot [y](k)$:

W.l.o.g. we assume $d_{x,y}(k) > 0$. Choose a clopen neighbourhood B of k, in which $[x]$ and $[y]$ vary less than 1. Then with $M := ([x](k)+1)([y](k)+1)$ for all clopen $C \subset B$ we have $M\,m(C) = ([x](k)+1)\,m(C)^{1/p} \cdot ([y](k)+1)\,m(C)^{1/p'} \geq \|\chi_C x\|\,\|\chi_C y\| \geq$

$\geq |\langle \chi_C x, \chi_C y\rangle| = |\langle \chi_C \chi_C x, y\rangle| = |\mu_{x,y}(C)| = |\int_C d_{x,y}\,dm|$.

Therefore $d_{x,y}(k) < \infty$ (otherwise the continuous function $d_{x,y}$ would be greater than M on a suitable clopen $C \subset B$ in contradiction to the above inequality). We choose a decreasing sequence (B_n) of clopen neighbourhoods of k, on which $[x]$, $[y]$, and $d_{x,y}$ vary less than $\frac{1}{n}$. Then as above for $\frac{1}{n} < d_{x,y}(k)$ we have

$[x](k)\cdot[y](k) + \frac{1}{n}([x](k) + [y](k) + \frac{1}{n}) \geq \frac{1}{m(B_n)}\int_{B_n} d_{x,y}\,dm \geq d_{x,y}(k) - \frac{1}{n}$.

Passing to the limit we get the required inequality.

II. By the above inequality and the bilinearity of the mapping $d_{\cdot,\cdot} : X \times Y \longrightarrow C^1(K;m)$ we get that $\langle x(k), y(k)\rangle_k := d_{x,y}(k)$ determines a well-defined bilinear mapping $\langle\cdot,\cdot\rangle_k : X_k \times Y_k \longrightarrow \mathbb{R}$.

<u>This mapping is faithful</u>:

As we have already the inequality of I., and the norms on Y_k and X_k' are absolutely homogeneous, it suffices to show that, given $y \in Y$ with $[y](k) = 1$ and $\varepsilon > 0$, there is an $x \in X$ with $[x](k) = 1$ and $d_{x,y}(k) \geq 1 - \varepsilon$. Then the canonical linear mapping from Y_k into X_k' is isometric and, by symmetry, we are done ($d_{x,y} = d_{y,x}$ because of $\langle fx,y\rangle = \langle x,fy\rangle$ for all $f \in C(K)$). By lemma 3.13 we may assume w.l.o.g. $[y] = \chi_B$ for a suitable clopen neighbourhood B of k with $m(B) < \infty$. We show the existence of an $x \in X$ with the desired (in fact stronger) properties in the following lemma:

<u>5.5 Lemma</u>: Let X, Y, K, $\langle\cdot,\cdot\rangle$ be as above. Then

for each $y \in Y$ with $[y] = \chi_B$, $\emptyset \neq B \subset K$, and $\varepsilon > 0$

there is an $x \in X$ with $[x] = \chi_B$ and $d_{x,y}|_B \geq 1 - \varepsilon$.

Proof of the lemma: We note that B is clopen and $0 < m(B) < \infty$, and apply the existence lemma 3.8 to

$A := \{x| \ x \in X, [x] = \chi_D, D \subset B, d_{x,y}|_D \geq 1 - \varepsilon\}$.

Condition (ii) of 3.8 is trivially satisfied, (iii) because of the continuity of $d_{x,y}$ and the fact that, for clopen $C \subset K$, $d_{E_C x,y}|_C = d_{x,y}|_C$. We give an indirect proof of (i). Assume $\emptyset \neq C \subset B$, C clopen, and that for no clopen D, $\emptyset \neq D \subset C$, there is an $x \in A$ with $[x] = \chi_D$. Then by continuity we have always $d_{x,y}|_D \leq 1 - \frac{\varepsilon}{2}$ for $D \subset C$ and x with $[x] = \chi_D$, hence $\langle x,y \rangle = \int_D d_{x,y}\,dm \leq (1 - \frac{\varepsilon}{2})\, m(D)$. In contradiction to this we find a $D_0 \subset C$, $x_0 \in X$ with $[x_0] = \chi_{D_0}$ and $\langle x_0,y \rangle > (1 - \frac{\varepsilon}{2})\, m(D_0)$:-

As the set $\mathcal{T}$ of steplike functions is dense in X (3. 14) and $\langle \cdot,\cdot \rangle$ is faithful, there is an $x \in \mathcal{T}$ with

$\langle x, \chi_C y \rangle > \|x\| \cdot \|\chi_C y\| (1 - \frac{\varepsilon}{2})$,

so $x = \sum_{\nu=1}^{n} \alpha_\nu x^\nu$ with $\alpha_\nu > 0$, $[x^\nu] = \chi_{D_\nu}$, $D_\nu \neq \emptyset$, pairwise disjoint, w.l.o.g. $D_\nu \subset C$ (because $\langle \chi_C x, \chi_C y \rangle = \langle x, \chi_C y \rangle$ and $\|\chi_C x\| \leq \|x\|$).

Let the quotient $k_\nu := \frac{|\langle x^\nu, y \rangle|}{m(D_\nu)}$ be maximal for $\nu = 1$ (w.l.o.g.).

Then $1 - \frac{\varepsilon}{2} < \frac{|\langle x, \chi_C y \rangle|}{\|x\| \cdot \|\chi_C y\|} \leq \frac{\Sigma \alpha_\nu |\langle x^\nu, y \rangle|}{(\Sigma \alpha_\nu{}^p m(D_\nu))^{1/p} (\Sigma\, m(D_\nu))^{1/p'}} \leq$

$$\leq \frac{\Sigma \alpha_\nu k_\nu m(D_\nu)}{\Sigma\, \alpha_\nu m(D_\nu)^{1/p} m(D_\nu)^{1/p'}} \leq \frac{k_1 \Sigma\, \alpha_\nu m(D_\nu)}{\Sigma\, \alpha_\nu m(D_\nu)} = \frac{|\langle x^1, y \rangle|}{m(D_1)}$$

, where we have to sum up $1 \leq \nu \leq n$ (for the third inequality we apply the Hölder inequality in the denominator).

$D_0 := D_1$ and $x_0 := \mathrm{sgn}\langle x^1, y\rangle \cdot x^1$ satisfy the desired inequality. With this all the conditions of the existence lemma are satisfied and there is an $x \in A$ with $[x] = \chi_B$, which proves the lemma and thus also the faithfulness of $\langle\cdot,\cdot\rangle_k$. □

B: Surjectivity and Reflexivity

Again let $1 < p < \infty$.

In contrast to the atomic case (with counting measure) the components Z_k of the integral module representing X' are weak-*-dense but, in general, proper subspaces of X_k', as example 5.9 will show. However, in some special cases we can ensure the surjectivity of the canonical isometries $Z_k \longrightarrow X_k'$.

5.6 Proposition: Let X, Y, K, $\langle\cdot,\cdot\rangle$ be as above. Then for $k \in K$

(i) $Z_k = X_k'$, if X_k is reflexive

(ii) X_k is reflexive if and only if Y_k is reflexive.

Proof: With X_k reflexive, we have $\sigma(X_k', X_k'') = \sigma(X_k', X_k)$ and norm closure and $\sigma(X_k', X_k)$ closure of convex sets coincide. So Z_k is weak-*-closed, hence $Z_k = X_k'$. So (i) is proved. With X_k reflexive, $Y_k \cong Z_k = X_k'$ is reflexive, too. The rest of (ii) follows from symmetry. □

Let k be an isolated point of K. One can easily verify that $Z_k = X_k'$, if Y is the dual X'. However, that does not hold for arbitrary pairs of integral modules, as one can see from the example $X = Y'$ with a non-reflexive space Y and trivial Boolean algebra $\mathfrak{U} = \{0, Id\}$.

If $X = \prod^p_{i\in I} X_i$, by the representation of X in 7.1 and the above remark we get the result $X' \cong \prod^{p'}_{i\in I} X_i'$ once again.

With the aid of theorem 5.4, one could attempt to characterize reflexivity of an integral module X by reflexivity of its components. As 5.6 shows, there is a close connection between the surjectivity of the canonical isometry $Z_k \longrightarrow X_k'$ and the reflexivity of X_k. We begin with

<u>5.7 Proposition</u>: Let X be a p-integral module in $\int_K^p X_k \, dm$.
If all components X_k $(k \in K)$ are reflexive, then X is also reflexive.

<u>Proof</u>: By 5.4 and 5.6 we know that X' has a representation $\sim$ as a p'-integral module in $\dot{\int_K^{p'}} X_k' dm$ and X" (as the dual of $\sim(X')$) has a representation $\approx$ as a p-integral module in $\int_K^p X_k'' dm$. We denote the natural embedding $x \longmapsto ev_x$ of X into X" by ev, the (surjective) embeddings of X_k into X_k'' by iso_k and put $iso_k(\infty) := \infty$. Evidently the restriction π of the product mapping

$$\mathop{\times}_{k \in K} iso_k : \mathop{\times}_{k \in K} X_k \,\dot{\cup}\{\infty\} \longrightarrow \mathop{\times}_{k \in K} X_k'' \,\dot{\cup}\{\infty\}$$

to X is an isometric C(K)-module isomorphism from X onto $\pi(X) \subset \int_K^p X_k'' dm$.
As the first step in the proof of the proposition we shall show that the following diagram commutes, i.e. $\pi = \approx \circ ev$:-

$$\begin{array}{ccccc}
X & = & X & \subset & \mathop{\times}_{k \in K} X_k \,\dot{\cup}\{\infty\} \\
\Big\downarrow ev & & \Big\downarrow \pi & & \Big\updownarrow \mathop{\times}_{k \in K} iso_k \\
X'' & \xrightarrow{\ \approx\ } & \int_K^p X_k'' dm & \subset & \mathop{\times}_{k \in K} X_k'' \,\dot{\cup}\{\infty\}
\end{array}$$

For this purpose let us denote the image of $f \in C(K)$ in $C_p(X'')$ by R_f instead of T_f (see ①). For all $x \in X$ and clopen $B \subset K$ we have, because $R_{\chi_B} = T_{\chi_B}''$ and $T''(ev_x) = ev_{Tx}$ for all $T \in [X]$, that

② $$\chi_B \widetilde{\widetilde{ev_x}} = \widetilde{\widetilde{R_{\chi_B} ev_x}} = \widetilde{\widetilde{T_{\chi_B}'' ev_x}} = \widetilde{\widetilde{ev_{T_{\chi_B} x}}} = \widetilde{\widetilde{ev_{\chi_B x}}} .$$

Therefore the isometric linear embedding $\approx \circ$ ev of X into $\approx(X'')$ commutes with characteristic projections (in fact it is a C(K)-module homomorphism, as the continuous step functions are dense in C(K)). From this and the continuity of the norm resolution it follows immediately that $[x] = [\widetilde{ev_x}]$ for all $x \in X$. In particular, we have

③ $$\widetilde{ev_x}(k) = \infty \text{ if and only if } x(k) = \infty \quad .$$

② shows that, for all $x \in X$, $\varphi \in X'$, and clopen $B \subset K$, $\mu_{\widetilde{ev_x},\tilde{\varphi}}(B) = \langle \chi_B \widetilde{ev_x}, \tilde{\varphi} \rangle = \langle \widetilde{ev_{\chi_B x}}, \tilde{\varphi} \rangle = \langle \chi_B x, \tilde{\varphi} \rangle = \mu_{x,\tilde{\varphi}}(B)$, where the first two bilinear forms are taken in the dual pair $(\approx(X''), \sim(X'))$, and the third in $(X, \sim(X'))$. Hence $\langle \widetilde{ev_x}(k), \tilde{\varphi}(k) \rangle_k = \langle x(k), \tilde{\varphi}(k) \rangle_k$ for $x(k) \neq \infty \neq \tilde{\varphi}(k)$, that means

④ $$\widetilde{ev_x}(k) = iso_k(x(k)) = \pi x(k) \quad \text{for } x(k) \neq \infty \quad .$$

③ and ④ yield $\pi = \approx \circ ev$.

Consequently the p-integral module $\pi(X)$ is a linear subspace of $\approx(X'')$. The maximality of p-integral modules (see 3.9) implies $\approx(ev(X)) = \pi(X) = \approx(X'')$. Because of the injectivity of ev this yields $ev(X) = X''$, which is the reflexivity of X. □

The converse of this proposition is false, as example 5.9 will show. However, we can describe the cases in which it holds.

<u>5.8 Proposition</u>: Let X be a reflexive p-integral module in $\int_K^p X_k \, dm$, $k \in K$. X_k is reflexive if and only if $Z_k = X_k'$.

<u>Proof</u>: "$\Longrightarrow$" has already been shown in 5.6. "$\Longleftarrow$": We shall apply the James reflexivity criterion (see [Di] or [J]) and show that each $\tilde{\varphi}(k) \in Z_k = X_k'$ attains its supremum on the unit ball of X_k.

Let $\varphi \in X'$ and $\tilde{\varphi}(k) \in Z_k$. W.l.o.g. $[\tilde{\varphi}](k) = 1$ and, by 3.13, $[\tilde{\varphi}] = \chi_B$

for a suitable clopen neighbourhood B of k. We apply the existence lemma to the set

$A := \{x \mid x \in X, \chi_D \le [x] = d_{x,\tilde{\varphi}} \le 2\chi_D$ for some clopen $D \subset B\}$.

If A satisfies the conditions (i), (ii), (iii) of 3.8, we obtain an $x \in A$ with $\chi_B \le [x] = d_{x,\tilde{\varphi}} \le 2\chi_B$, in particular $0 \ne x(k) \ne \infty$ and $\tilde{\varphi}(k)(x(k)) = d_{x,\tilde{\varphi}}(k) = [x](k)$, and we are done.

Again (ii) is trivially satisfied, (iii) by continuity. We prove (i). Let C be a clopen, non-empty subset of B. As X is reflexive, there is a $z \in X$ with $|S\chi_C\varphi(z)| = \|S\chi_C\varphi\| \cdot \|z\| \ne 0$, and therefore $\int d_{z,\chi_C\tilde{\varphi}}\,dm =$ $= |S_{\chi_C}\varphi(z)| = (\int[\chi_C\tilde{\varphi}]^{p'}\,dm)^{1/p'}(\int[z]^p\,dm)^{1/p} \ge \int[\chi_C\tilde{\varphi}][z]\,dm =$ $= \int[\chi_C z]\,dm$. Both integrands are continuous and $d_{z,\chi_C\tilde{\varphi}} = d_{\chi_C z,\tilde{\varphi}} \le$ $\le [\tilde{\varphi}]\cdot[\chi_C z] = [\chi_C z] \ne 0$. Therefore $d_{\chi_C z,\tilde{\varphi}} = [\chi_C z] \ne 0$.

W.l.o.g. $[z](j) = \frac{3}{2}$ for a suitable $j \in C$, and we find a clopen neighbourhood D of j in C with $\chi_D \le [x] = d_{x,\tilde{\varphi}} \le 2\chi_D$, where $x := \chi_D z$. □

The following example of an atomic integral module shows that not all components of a reflexive integral module need be reflexive. Consequently, by the preceding proposition, there are components X_k with $Z_k \subsetneqq X_k'$.

5.9 Let l_n^∞ be $\mathbb{R}^n$ with the supremum norm. The Banach space $\Pi^p_{n\in\mathbb{N}} l_n^\infty$ is reflexive as a p-product of reflexive spaces. It has a representation as a p-integral module on $K = \beta\mathbb{N}$ with the measure counting the elements of $\mathbb{N}$ and the components l_n^∞ for $n \in \mathbb{N}$ (7.1).

It has another representation $\sim$ as a p-integral module X in $\int^p_{\beta\mathbb{N}} X_k\,dm$, where m is the perfect Borel measure on $\beta\mathbb{N}$ defined by $m(\{n\}) := 2^{-n}$

for $n \in \mathbb{N}$. Because of 3.12 we may assume w.l.o.g. $X_n = l^\infty_n$ and for the image $\tilde{x}$ in X of $x = (x_n) \in \prod^p_{n\in\mathbb{N}} l^\infty_n$ we get $\tilde{x}(n) = 2^n x_n$ $(n \in \mathbb{N})$. Let us denote the canonical projection from l^∞ onto l^∞_n by p_n, i.e. $p_n((\alpha_\nu)_{\nu\in\mathbb{N}}) = (\alpha_\nu)_{\nu\in\{1,..,n\}}$.

Let $k \in \beta\mathbb{N}\setminus\mathbb{N}$. We shall embed l^∞ isometrically into X_k.

For $\alpha \in l^\infty$ we have $\alpha^* := (2^{-n} p_n(\alpha)) \in \prod^p_{n\in\mathbb{N}} l^\infty_n$ because of $\|p_n(\alpha)\| \le \|\alpha\|$. As $[\widetilde{\alpha^*}](n) = \|2^n 2^{-n} p_n(\alpha)\| = \|p_n(\alpha)\| \xrightarrow[n\to\infty]{} \|\alpha\|$ and k is a cluster point of each subset of $\mathbb{N}$ with finite complement (in $\mathbb{N}$), we have $[\widetilde{\alpha^*}](k) = \|\alpha\| \neq \infty$. Therefore $\Phi_k : l^\infty \longrightarrow X_k$, defined by $\alpha \longmapsto \Phi_k(\alpha) := \widetilde{\alpha^*}(k)$, is an isometric and obviously linear mapping. X_k cannot be reflexive, because the closed subspace $\Phi_k(l^\infty)$ is not reflexive.

We note that in the preceding example X_k is not reflexive exactly for the k of $\beta\mathbb{N}\setminus\mathbb{N}$. This suggests that an integral module is reflexive if and only if m-almost all components are reflexive. In fact we can improve proposition 5.7 :-

<u>5.1o Proposition</u>: Let X be a p-integral module in $\int^p_K X_k \, dm$.
If m-almost all components X_k are reflexive, then X is also reflexive.

<u>Proof</u>: It suffices to show that ev(X) is norm dense in X". Let $f \in X''$, $\varepsilon > 0$. $N := \{k \mid X_k \text{ not reflexive}\}$ is of first category, hence $N = \bigcup_{\nu\in\mathbb{N}} A_\nu$ with nowhere dense A_ν $(\nu \in \mathbb{N})$. For each $\nu \in \mathbb{N}$ the clopen subsets of K including A_ν form a decreasing net whose infimum (interior of the intersection) is the empty set. Because of 1.6 (ii) there are clopen $C_\nu \supset A_\nu$ with $\| \chi_{C_\nu} \tilde{f} \|^p \le \varepsilon \cdot 2^{-\nu}$, where $\approx$ is the representation of X" in 5.7. With $C := (\bigcup_{\nu\in\mathbb{N}} C_\nu)^-$ we have $N \subset C$ and (because of

1.6 (i)) $\|R_{\chi_C} f\|^p = \|\chi_C \tilde{f}\|^p \leq \varepsilon$.

$\chi_{K\setminus C}X$ is a p-integral module whose components are reflexive, hence itself reflexive by 5.7 . We have an obvious isometry between $(\chi_{K\setminus C}X)''$ and $R_{\chi_{K\setminus C}}(X'')$; therefore there is an $x \in \chi_{K\setminus C}X$ with $ev_x = R_{\chi_{K\setminus C}} f$. For this $x \in X$ we have $\|f - ev_x\|^p = \|R_{\chi_C} f\|^p \leq \varepsilon$.□

Whether the converse of this proposition holds, is unknown (i.e., it is unknown whether $Z_k = X_k'$ for almost all components X_k of a reflexive p-integral module in $\int_K^p X_k\, dm$). There seems to be a connection between this problem and that of constructing a p'-integral module in $\int_K^{p'} Y_k\, dm$, which contains a given p'-integral module in $\int_K^{p'} Z_k\, dm$ with $Z_k \subset Y_k$ for all $k \in K$.

Chapter 6: Spectral Theory for L^p-Operators

The aim of this chapter is to represent the elements of the Cunningham p- algebra $C_p(X)$, which we shall call L^p-operators, as Stieltjes integrals over families of L^p-projections (spectral families) and to describe their properties by means of those of the corresponding spectral families. The theory is analogous to the corresponding theory for self-adjoint operators in Hilbert space.

In what follows let X be a fixed real Banach space, $1 \leq p < \infty, p \neq 2$.

For $E \in \mathbb{P}_p$ let $J_E : = E(X)$, $B_E \subset \Omega_p$ the corresponding clopen set; for $T \in C_p(X)$ we write $\hat{T} : = \psi(T)$ in $C(\Omega_p)$, for example $\hat{E} = \chi_{B_E}$.

6.1 Definition:

(i) A family $(E_\lambda)_{\lambda \in R}$ of L^p-projections is called a spectral family if $\lambda \leq \mu$ implies $E_\lambda \leq E_\mu$.

(ii) A spectral family has a bounded support if there exist $m, M \in R$ such that $\lambda < m$ implies $E_\lambda = 0$ and $\lambda \geq M$ implies $E_\lambda = Id$.

(iii) A spectral family is called normalized if for every $\lambda_o \in R$ we have $E_{\lambda_o} = \inf\limits_{\lambda > \lambda_o} E_\lambda$.

Note: In (i) and (ii) the corresponding families $(J_\lambda)_{\lambda \in R}$ and $(B_\lambda)_{\lambda \in R}$ ($J_\lambda : = J_{E_\lambda}$, $B_\lambda : = B_{E_\lambda}$) have similar monotonic properties. In future all $(E_\lambda)_{\lambda \in R}$ have bounded support.

6.2 Lemma: For any spectral family $(E_\lambda)_{\lambda \in R}$, the following are equivalent:

(i) $(E_\lambda)_{\lambda \in R}$ is normalized

(ii) $J_{\lambda_o} = \bigcap\limits_{\lambda > \lambda_o} J_\lambda$ for every $\lambda_o \in R$

(iii) $B_{\lambda_o} = (\bigcap_{\lambda>\lambda_o} B_\lambda)^o = \inf_{\lambda>\lambda_o} B_\lambda$ for every $\lambda_o \in R$

(iv) for every $\lambda_o \in R$, $x \in X$ and every real sequence $(\lambda_n)_{n \in \mathbb{N}}$

with $\lambda_n > \lambda_o$, $\lambda_n \to \lambda_o$: $\lim_{n\to\infty} E_{\lambda_n} x = E_{\lambda_o} x$.

Proof: The equivalence is a consequence of the fact that the Boolean algebras involved are isomorphic and that for decreasing families the infimum of L^p-projections is the limit in the strong operator topology. □

The following technical lemma is of fundamental importance for what follows:

6.3 Lemma: Let $E_1 \le E_2 \le \ldots \le E_n \in P_p$, $x \in X$, $a_1,\ldots,a_n \in R$. Then

$$\| \sum_{i=1}^{n-1} a_i (E_{i+1} - E_i)x - a_n (E_n - E_1)x \|^p \le$$

$$\le \max_{1\le i\le n-1} |a_i - a_n|^p \, \|(E_n - E_1)x\|^p$$

Proof.: We obtain for $b_1,\ldots,b_k \in R$:

$$\| \sum_{i=1}^{k} b_i(E_{i+1} - E_i)x \|^p = \sum_{i=1}^{k} |b_i|^p \cdot \|(E_{i+1} - E_i)x\|^p ,$$

and so $\| \sum_{i=1}^{n-1} a_i(E_{i+1}-E_i)x - a_n(E_n - E_1)x\|^p = \|\sum_{i=1}^{n-1} (a_i-a_n)(E_{i+1}-E_i)x\|^p$

$$= \sum_{i=1}^{n-1} |a_i - a_n|^p \, \|(E_{i+1} - E_i)x\|^p \le \max_{1\le i\le n-1} |a_i-a_n|^p \sum_{i=1}^{n-1} \|(E_{i+1}-E_i)x\|^p$$

$$= \max_{1\le i\le n-1} |a_i - a_n|^p \, \|(E_n - E_1)x\|^p .$$ □

For a compact interval $[a,b] \subset R$ we examine now the system $\underline{Z}$ of all partitions Z of $[a,b]$ together with intermediate points, i.e.

$Z := \{t_o,t_1,\ldots,t_n, \tau_o,\tau_1,\ldots,\tau_{n-1}\}$ with

$a = t_o < t_1 < \ldots < t_n = b$ and $t_\nu \le \tau_\nu \le t_{\nu+1}$, $\nu = 0,\ldots,n-1$.

We write $Z = (t_o \le \tau_o \le \ldots \le \tau_{n-1} \le t_n)$

and further $\delta(Z) := \max_{\nu=0,\ldots,n-1} |t_{\nu+1} - t_\nu|$.

A sequence of partitions $(Z_k)_{k \in \mathbb{N}}$ is called <u>distinguished</u> (d.s.p.) if $\lim_{k\to\infty} \delta(Z_k) = 0$.

For Z and $Z' := (t'_0 \le \tau'_0 \le t'_1 \le \dots \le \tau'_{m-1} \le t'_m) \in \underline{Z}$ we define the <u>refinement</u> $Z^\vee := Z \vee Z'$ to be that partition in which all the t_ν, t'_μ appear in their natural order as $t^\vee_\rho$, $\rho = 0,\dots 1$ with intermediate points $\tau^\vee_\rho := \frac{t^\vee_{\rho+1} + t^\vee_\rho}{2}$ (if a t_ν and a t'_μ are equal only one should be used in the refinement); clearly $\delta(Z^\vee) \le \min\{\delta(Z), \delta(Z')\}$.

<u>6.4 Definition</u>: Let $(E_\lambda)_{\lambda \in R}$ be a spectral family with $a < m \le M < b$, $\varphi : [a,b] \to R$ continuous, $Z \in \underline{Z}$, $x \in X$. Then $\Sigma(\varphi, Z, x) := \sum_{i=0}^{n-1} \varphi(\tau_i)(E_{t_{i+1}} - E_{t_i})x$.

<u>6.5 Proposition</u>: With the definition of 6.4 and for a d.s.p., $(\Sigma(\varphi, Z_m, \cdot))_{m \in \mathbb{N}}$ converges in $[X]$ and the limit is in $C_p(X)$. It is independent of the special choice of $(Z_m)_{m\in\mathbb{N}}$ and a,b and will be denoted by $\int \varphi(\lambda) dE_\lambda$.

<u>Proof</u>: The proof is in two stages: First we show for $Z, Z' \in \underline{Z}$

$$\| \Sigma(\varphi, Z, x) - \Sigma(\varphi, Z \vee Z', x) \|^p \le (\max_{i=0,\dots,n-1} \eta_i(Z)^p) \| x \|^p,$$

whereby $\eta_i(Z) := \max_{t_i \le s, t \le t_{i+1}} | \varphi(t) - \varphi(s)|$ is the variation of φ on $[t_i, t_{i+1}]$. To this end we write $\Sigma(\varphi, Z \vee Z', x)$ as a double sum over the t_i of Z and the points between t_i and t_{i+1} and apply 6.3. Thus we obtain $\| \Sigma(\varphi, Z_l, x) - \Sigma(\varphi, Z_k, x)\| \le (\max \eta_i(Z_l) + \max \eta_j(Z_k))\|x\|$. Therefore the sequence $(\Sigma(\varphi, Z_m, \cdot))_{m\in\mathbb{N}}$ is Cauchy by the uniform continuity of φ. Its uniform limit $\int \varphi(\lambda) dE_\lambda$ is in $C_p(X)$. The independence from a,b is clear by 6.1(ii), and it becomes independent of the special $(Z_m)_{m\in\mathbb{N}}$ by convergence of mixed sequences. □

Now we construct for a given $T \in C_p(X)$ a spectral family $(E_\lambda)_{\lambda \in R}$ such that $T = \int \lambda \, dE_\lambda$:-

<u>6.6. Definition</u>: If $T \in C_p(X)$, $\lambda \in R$, then $B_\lambda := \{\omega | \omega \in \Omega_p, \hat{T}(\omega) \le \lambda\}^\circ$.

Because $\hat{T}$ is continuous and Ω extremally disconnected, B_λ must be clopen, and the corresponding E_λ's form an - in fact normalized - spectral family; moreover we obtain

<u>6.7 Theorem</u>: $T = \int \lambda \, dE_\lambda$.

<u>Proof</u>: There exist, for given $\epsilon > 0$, pairwise disjoint $E_1, \ldots, E_n \in \mathbb{P}_p$ and $a_1 < \ldots < a_n \in R$ with $\sum_{i=1}^{n} E_i = Id$ such that

$$\| T - \sum_{i=1}^{n} a_i E_i \| = \| \hat{T} - \sum_{i=1}^{n} a_i \hat{E}_i \| < \epsilon .$$

For $i_0 \in \{1, \ldots, n\}$ and $x \in E_{i_0}(X)$ it follows that

$E_\mu x = x$ if $\mu \geq a_{i_0} + \epsilon$ and $E_\lambda x = o$ if $\lambda \leq a_{i_0} - \epsilon$. According to 6.3, for $Z \in \underline{Z}$ with, w.l.o.g., $a_{i_0} + \epsilon$, $a_{i_0} - \epsilon$ as partition points, we have $\|a_{i_0} x - \Sigma(Id_R, Z, x)\|^p \leq$

$$\leq \max_{a_{i_0} - \epsilon \leq \tau_i \leq a_{i_0} + \epsilon} |a_{i_0} - \tau_i|^p \cdot \|(E_{a_{i_0} + \epsilon} - E_{a_{i_0} - \epsilon})x\|^p \leq \epsilon^p \cdot \|x\|^p .$$

Passing to the limit $S := \int \lambda \, dE_\lambda$ it follows that

$\epsilon\|x\| \geq \|a_{i_0} x - Sx\|$. But $\sum_{i=1}^{n} a_i E_i x = a_{i_0} x$, therefore

$\| (S - T)x \| \leq 2 \epsilon \cdot \|x\|$. For any $x \in X$ we have

$\| (S - T)x\|^p = \| \Sigma(S - T)E_i x\|^p = \Sigma \| (S - T)E_i x \|^p \leq$

$\leq 2^p \epsilon^p \Sigma \|E_i x \|^p = 2^p \epsilon^p \|x\|^p$, that is $\| S - T \| \leq 2 \epsilon$.

Hence $S = T$, because $\epsilon > 0$ is arbitrary. □

The uniqueness question is answered by the following proposition: -

<u>6.8 Proposition</u>: For any normalized spectral family $(E'_\lambda)_{\lambda \in R}$ with

$T = \int \lambda \, dE'_\lambda$

$$E_\lambda = E'_\lambda \quad \text{f.a. } \lambda \in R.$$

<u>Proof</u>: The steps of the proof are

(i) For given $(E'_\lambda)_{\lambda \in R}$ with support in $[m', M]$ where $0 < m'$

we have $\underline{O} \leq \hat{T} \leq \underline{M}$

(ii) for $0 \leq_p T$, $E_\lambda = E'_\lambda$

(iii) the statement holds for every $(E'_\lambda)_{\lambda \in R}$

(i) let $m' > 0$; then, for $\varepsilon > 0$, there exists a partition $Z = (t_0 \leq \tau_0 \leq t_1 \leq \ldots \leq \tau_{n-1} \leq t_n)$ with $\tau_i \geq t_i \geq m > 0$ such that $\|T - \sum_{i=0}^{n-1} \tau_i (E'_{t_{i+1}} - E'_{t_i})\| < \varepsilon$. It follows that $T \in C_p(X)^+$ (see 2.3). Otherwise, for any partition $\tilde{Z}$ of $[0,M+\varepsilon]$ and any $x \in X$ $\| \sum \tilde{\tau}_i (E_{\tilde{t}_{i+1}} - E_{\tilde{t}_i})x\|^p \leq (M + \varepsilon)^p \|x\|^p$, thus $\|Tx\|^p \leq (M + \varepsilon)^p \|x\|^p$, and $\|Tx\| \leq M \|x\|$, so that $T \leq_p M$ Id, and finally $\underline{O} \leq \hat{T} \leq \underline{M}$.

(ii) : for $m' > 0$, $E'_{\lambda_0} = 0 = E_{\lambda_0}$ f.a. $\lambda_0 < 0$, because by (i) B_{λ_0} must be the empty set. Let $\lambda_0 \geq 0$, B'_{λ_0} the corresponding clopen set. We show $B_{\lambda_0} = B'_{\lambda_0}$: " $\subseteq$ " : let $x \in E'_{\lambda_0}(X)$, Z a partition of $[0,b]$ with λ_0 as a partition point. Then by 6.3 we get

$$\| \sum (Id_R, Z, x)\|^p = \| \sum_{t_{i+1} \leq \lambda_0} \tau_i (E'_{t_{i+1}} - E'_{t_i})x\|^p \leq$$

$$\leq \lambda_0^p \sum_{t_{i+1} \leq \lambda_0} \| (E'_{t_{i+1}} - E'_{t_i})x\|^p = \lambda_0^p \|x\|^p ,$$

hence $\|Tx\| \leq \lambda_0 \|x\|$, so $\underline{O} \leq \hat{T} \cdot \hat{E}'_{\lambda_0} \leq \lambda_0 \hat{E}'_{\lambda_0}$, therefore $\hat{T}(\omega) \leq \lambda_0$ for $\omega \in B'_{\lambda_0}$, and so $B_{\lambda_0} \subseteq B'_{\lambda_0}$. Similarly we can show indirectly that $B_{\lambda_0} \subseteq B'_{\lambda_{0+\varepsilon}}$, which gives $B_{\lambda_0} \subseteq B'_{\lambda_0}$ by the normality of $(E'_\lambda)_{\lambda \in R}$.

(iii) can be reduced to cases (i)-(ii) by setting $E''_\lambda := E'_{\lambda+a} (a<m)$. □

<u>6.9 Corollary</u>: (a) for two normalized $(E_\lambda)_{\lambda \in R}$, $(E'_\lambda)_{\lambda \in R}$ the following are equivalent: -

(i) $(E_\lambda)_{\lambda \in R} = (E'_\lambda)_{\lambda \in R}$

(ii) $\int \lambda \, dE_\lambda = \int \lambda \, d E'_\lambda$

(b) there is a one-to-one correspondence between the elements of $C_p(X)$ and the normalized spectral families. □

<u>6.10 Definition</u>: For $T \in C_p(X)$ let the spectra $\sigma(T), \sigma_p(T) \subset \mathbb{R}$ be defined by

$\lambda \notin \sigma(T)$ iff there is $S \in [X]$ such that $S \circ (\lambda Id - T) = (\lambda Id - T) \circ S = Id$

$\lambda \notin \sigma_p(T)$ iff there is $S \in C_p(X)$ such that $S \circ (\lambda Id - T) = (\lambda Id - T) \circ S = Id$.

<u>6.11 Proposition</u>: Let $T \in C_p(X)$, $J_o = E_o(X)$ the L^p-summand corresponding to $B_o := \{\omega | \omega \in \Omega_p, \hat{T}(\omega) = 0\}^o$. Then $\ker T = J_o$; in particular, $\ker T$ is an L^p-summand.

<u>Proof</u>: We use the representation of X as a p-integral module and show the equivalent statement

$\hat{T}[x] = 0 \Leftrightarrow \chi_{B_o} [x] = [x]$

" $\Leftarrow$ " is clear; " $\Rightarrow$ " : for $\omega \in B_o$, $[x](\omega) = \chi_{B_o} [x](\omega)$. But for $\omega \notin B_o$, $[x](\omega) = 0 = \chi_{B_o} [x](\omega)$ too, since $[x]$ vanishes on $\Omega_p \setminus \hat{T}^{-1}(0)$, and then also on $\Omega_p \setminus B_o$ because $[x]^{-1}(0)$ is closed. □

<u>6.12 Corollary</u>: (a) Every L^p-operator T maps every L^p-summand $J = E[X]$ into itself, and $T^{-1}(J)$ is also an L^p-summand

(b) If for $\lambda \in \mathbb{R}$, $U_\lambda := \ker(\lambda Id - T)$ denotes the corresponding eigenspace, then U_λ is an L^p-summand.

<u>Proof</u>: The proof is clear since $T \circ E = E \circ T = E^2 \circ T = E \circ T \circ E$ and $T^{-1}(J) = \ker(T \circ E - E)$. □

<u>6.13 Lemma</u>: Let $(E_\lambda)_{\lambda \in \mathbb{R}}$ be a spectral family, $\varphi, \psi: [m,M] \to \mathbb{R}$ continuous, $\alpha \in \mathbb{R}$. Then

(i) $(\int \varphi(\lambda) dE_\lambda) \dot{+} (\int \psi(\lambda) dE_\lambda) = \int (\varphi \dot{+} \psi)(\lambda) dE_\lambda$

(ii) $\int \alpha \, dE_\lambda = \alpha \, Id$

(iii) if $(E_\lambda)_{\lambda \in R}$ is constant on $[\lambda_o - \varepsilon , \lambda_o + \varepsilon]$ and $\varphi(\lambda) = \psi(\lambda)$ for $|\lambda - \lambda_o| \geq \varepsilon$, then

$$\int \varphi(\lambda) dE_\lambda = \int \psi(\lambda) dE_\lambda$$

Proof: We obtain (i),(ii) by proving the corresponding formulas for sums over a partition Z and passing to the limit (the operations are continuous); similarly we get (iii), if $\lambda_o - \varepsilon$, $\lambda_o + \varepsilon$ are partition points of Z. □

Now we can characterize the spectrum of an L^p-operator:

6.14 Proposition: Let $T \in C_p(X)$,$(E_\lambda)_{\lambda \in R}$ the corresponding normalized spectral family.

I. The following are equivalent: -

(i) $\lambda_o \notin \sigma(T)$

(ii) $\lambda_o \notin \sigma_p(T)$

(iii) $(E_\lambda)_{\lambda \in R}$ is constant in a neighbourhood of λ_o

(iv) $\lambda_o \notin \hat{T}(\Omega_p)$

II. For $\lambda_o \in R$ the eigenspace U_{λ_o} is the discontinuity at λ_o in the family $(J_\lambda)_{\lambda \in R}$ corresponding to $(E_\lambda)_{\lambda \in R}$:

$$(\bigcup_{\lambda < \lambda_o} J_\lambda)^- \oplus_p U_{\lambda_o} = J_{\lambda_o} .$$

Proof:

I, (i) ⇒ (iv) : If $\lambda_o \in \hat{T}(\Omega_p)$ then for all

$\varepsilon > 0$ $\quad \{\omega | \omega \in \Omega_p, |\hat{T} \omega - \lambda_o| \leq \varepsilon\}^o =: B_\varepsilon \neq \emptyset$ so for x with $\|x\| = 1$, $x \in$ range E_{B_ε}, $\| (T - \lambda_o \, Id)x\| \leq \varepsilon$, a contradiction.

(iv) ⇒ (iii): there exists $\varepsilon_o > 0$ such that range $\hat{T} \cap [\lambda_o - \varepsilon_o, \lambda_o + \varepsilon] = \emptyset$, hence $(E_\lambda)_{\lambda \in R}$ is constant for $|\lambda - \lambda_o| \leq \varepsilon_o$

(iii) ⇒ (ii) : if $(E_\lambda)_{\lambda\in R}$ is constant for $|\lambda - \lambda_0| < \varepsilon_0$, then we choose $\varphi : R \to R$ continuous such that $\varphi(\lambda) = (\lambda-\lambda_0)^{-1}$ for $|\lambda-\lambda_0| \geq \varepsilon$. By 6.13 we get that $S := \int \varphi(\lambda) dE_\lambda$ is an inverse to $(T - \lambda_0 \, Id)$ in $C_p(X)$.

(ii) ⇒ (i) is trivial, (ii) ⇔ (iv) follows directly from

$$C_p(X) \cong C(\Omega_p).$$

II. is obtained by verifying the equivalent relation

$$[\bigcup_{\lambda<\lambda_0} \{\hat{T} \leq \lambda\}^0]^- \dot{\cup} \{\hat{T} = \lambda_0\}^0 = \{\hat{T} \leq \lambda_0\}^0 \text{ in } \Omega_p . \qquad \square$$

<u>Corollary 6.15</u> (a) For $T \in C_p(X)$ with corresponding $(E_\lambda)_{\lambda\in R}$ the set $\sigma_p(T) = \sigma(T)$ consists of those $\lambda_0 \in R$ for which $(E_\lambda)_{\lambda\in R}$ is not constant in any neighbourhood.

(b) If S^{-1} exists for $S \in [X]$ then $S^{-1} \in C_p(X)$ iff $S \in C_p(X)$.

(c) For $\lambda_0 \in R$ $U_{\lambda_0} \neq 0$ iff $(E_\lambda)_{\lambda\in R}$ is discontinuous at λ_0 .

(d) If $\lambda_0 \in \sigma(T)$ is isolated, then λ_0 is an eigenvalue. □

In conclusion we prove some propositions e.g. the "spectral mapping theorem" 6.19, for which we need some further properties of the Stieltjes integrals $\int \varphi(\lambda) dE_\lambda$:

<u>6.16 Lemma</u>: Let $T \in C_p(X), (E_\lambda)_{\lambda\in R}$ the corresponding spectral family, $\varphi , \psi : R \to R$ continuous. Then

(i) $\varphi_{|\sigma(T)} = 0 \Leftrightarrow \int \varphi(\lambda) dE_\lambda = 0$

(ii) $\varphi_{|\sigma(T)} = \psi_{|\sigma(T)} \Leftrightarrow \int \varphi(\lambda) dE_\lambda = \int \psi(\lambda) dE_\lambda$

(iii) $\| \int \varphi(\lambda) dE_\lambda \| = \| \varphi_{|\sigma(T)} \|$

Proof:

(i) " $\Rightarrow$ " : For given $\varepsilon > 0$ there exist, by compactness, $\bar{s}_1 < \bar{t}_1 < \bar{s}_2 < \ldots < \bar{s}_m < \bar{t}_m$ such that $\sigma(T) \subset \bigcup_{j=1}^{m}]\bar{s}_j, \bar{t}_j[$, and $|\varphi(\tau)| \le \varepsilon$ for $\tau \in \bigcup_{j=1}^{m}]\bar{s}_j, \bar{t}_j[$. If we now examine a partition Z which contains all the $\bar{s}_j$'s and $\bar{t}_j$'s as partition points, we get, as in 6.5, $\| \sum(\varphi,Z,x)\|^p \le \varepsilon^p \|x\|^p$ for all $x \in X$, such that $\int \varphi(\lambda)dE_\lambda = 0$.

" $\Leftarrow$ " : If $\varphi_{|\sigma(T)} \neq 0$, then $|\varphi|_{|[\lambda_0 - \varepsilon, \lambda_0 + \varepsilon]} \ge \delta > 0$ for suitable λ_0 and $\varepsilon > 0$. For a partition Z with $\lambda_0 - \varepsilon, \lambda_0 + \varepsilon$ as partition points we get $\| \sum(\varphi,Z,x)\|^p =$

$$= \sum_{\lambda_0 - \varepsilon \le t_i < \lambda_0 + \varepsilon} |\varphi(\tau_i)|^p \, \|(E_{t_{i+1}} - E_{t_i})x\|^p \ge \delta^p \, \|x\|^p \text{ for all } x \in X$$

and therefore $\| \int \varphi(\lambda)dE_\lambda \| \neq 0$.

(ii) follows from (i) because of the linearity of the integral. For the proof of (iii), " $\le$ " we note that for any partition Z and $x \in X$ $\|(\varphi,Z,x)\|^p = \sum_{i=0}^{n} |\varphi(\tau_i)|^p \cdot (\|E_{t_{i+1}} x\|^p - \|E_{t_i} x\|^p)$, so that

$$\|(\int \varphi(\lambda)dE_\lambda)x\|^p = \int |\varphi(\lambda)|^p \, d\, \|E_\lambda x\|^p \le \|\varphi\|^p \cdot \|x\|^p ;$$

" $\ge$ " follows as in (i), " $\Leftarrow$ " . □

6.17 Proposition: Let $T \in C_p(X)$, $A(T)$ the Banach algebra generated by T and Id. Then $\Psi_T : C(\sigma(T)) \to A(T)$

$$\varphi \mapsto \int \varphi(\lambda)dE_\lambda$$

is an isometric order isomorphism .

Proof: By 6.16 and 6.13, Ψ_T is an isometric algebra homomorphism taking values in $C_p(X)$. The images of polynomials are polynomials in $A(T)$ (6.13), hence range $\Psi_T = A(T)$ (Stone-Weierstraß); Ψ_T is also positive, because the ordering can be expressed by means of the norm ($\varphi \geq 0 \Leftrightarrow \|\varphi - \|\varphi\|\,\| \leq \|\varphi\|$). □

With the natural <u>definition</u> $\varphi(T) := \psi_T(\varphi)$ we obtain the following results which can be proved by applying continuity arguments and the Weierstraß approximation theorem.

<u>6.18 Proposition</u>: For $\varphi \in C(\sigma(T))$, $\varphi(T) = \varphi \circ \hat{T}$ in $C(\Omega_p)$ (see 6.14). □

<u>6.19 Proposition</u>: $\sigma(\varphi(T)) = \varphi(\sigma(T))$. □

<u>6.20 Proposition</u>: For $\psi \in C(\sigma(\varphi(T))$,

$$\psi(\varphi(T)) = (\psi \circ \varphi)(T).$$ □

<u>6.21 Proposition</u>: The following are equivalent

(i) $\dim A(T) < \infty$

(ii) $\sigma(T)$ is finite

(iii) T is a finite linear combination of L^p-projections

(iv) There is a polynomial P with $P(T) = 0$. □

Chapter 7: The L^p-structure of the Bochner spaces and related results

In chapter 3 we showed how to represent any Banach space X as a kind of vector-valued L^p-space with respect to any complete Boolean algebra $\mathfrak{A}$ of L^p-projections. In this chapter we look at how this representation behaves in the case of the spaces $\prod_{i\in I}{}^p X_i$ and the Bochner spaces $L^p(\mu;V)$ which are themselves vector-valued L^p-spaces - though in a different sense. We go on to examine some similarities in the behaviour of general integral modules and those which represent Bochner spaces.

A: Product spaces

For the definition of the spaces $\prod_{i\in I}{}^p X_i$ see chapter 2. We naturally consider the case $p<\infty$. These spaces are the natural vector-valued generalizations of the spaces l^p_I. Although there may well be L^p-projections in the component spaces X_i, there is a natural complete Boolean algebra of L^p-projections on $\prod_{i\in I}{}^p X_i$ which depends only on I (as long as the X_i's are non-zero), namely that generated by the canonical projections onto the components (2.6). We call this algebra $\mathfrak{A}$.

7.1 Proposition: With the above notation

(i) The Stonean space of $\mathfrak{A}$ is homeomorphic to βI

(ii) If m is the perfect measure on βI which counts the points of I, and $Y \subset \int_{\beta I}^{p} Y_k dm$ an integral module representation of $\prod_{i\in I}{}^p X_i$ then $Y_k \cong \{0\}$ for $k \in \beta I \setminus I$ and $Y_i \cong X_i$ for $i \in I$.

Proof: (i) $\mathfrak{A}$ is clearly the algebra of all $E_J:(x_i) \mapsto (\chi_J(i)x_i)$

where J runs through the subsets of I. The correspondence $E_J \leftrightarrow J^-$ (closure in βI) shows that the Stonean space of this algebra is βI.

(ii) Since every neighbourhood of a point k in $\beta I \setminus I$ contains infinitely many points of I, k is an m-null point and therefore $X_k = \{0\}$. On the other hand, if $j \in I$, the projection $E_{\{j\}}$ in $\mathfrak{A}$ corresponds to the application of $\chi_{\{j\}}$ to Y (since $\{j\} = \{j\}^-$). Since the representation respects the action of the projections in $\mathfrak{A}$, the ranges are isometric. The two range spaces are identifiable with X_j and Y_j respectively so that $X_j \cong Y_j$. □

This short proposition shows that the integral module representation with respect to this m is merely another way of writing the natural representation as a sum indexed by I. If we had used another perfect measure instead of m this would not necessarily have been the case, and since we asserted earlier that any perfect measure is as good as any other there would seem to be no reason for making this particular choice (other than the fact that it leads to Proposition 7.1(ii)). However, the fact that the points of I are all isolated in βI means that they cannot be m-null points for any perfect measure m. The measure which counts the points of I is thus maximal in the sense that no other perfect measure has more null points. Proposition 3.12 shows that proposition 7.1 (ii) holds for any perfect measure which is maximal in this sense. We have seen (cf. the counter-example in chapter 5) that it is nevertheless useful to study other representations since we can then infer properties of the non-zero component spaces in the compactification from properties of the known spaces

X_i (see also section C of this chapter).

B: The Bochner spaces $L^p(\mu;V)$

7.2 Definition: If μ is a measure on some measure space and V a Banach space, we denote (for $1 \leq p < \infty$) by $L^p(\mu;V)$ the space of equivalence classes of V-valued strongly μ-measurable functions f for which $\|f\|_p := (\int \|f(k)\|_V^p d\mu)^{1/p} < \infty$ (cf. [H Ph], p. 46).

These spaces are the simplest examples of spaces of vector-valued integrable functions over arbitrary measure spaces. Although there may of course be L^p-projections on the space V there is, for fixed μ, an algebra $\mathfrak{A}$ of L^p-projections which is independent of the structure of V, namely that comprising the projections E_B: $f \mapsto \chi_B f$ for μ-measurable B. If Ω is the Stonean space of this algebra there is a natural perfect measure m on Ω which is induced by μ. Each clopen set D in Ω corresponds to a projection E_B in $\mathfrak{A}$. Although B is not uniquely determined by this, its measure is, so that we can define a perfect measure m by means of m(D) := $\mu(B)$ where D runs through the set of all clopen subsets of Ω.

7.3 Proposition: If $X \subset \int_\Omega^p X_k dm$ is an integral module representation of $L^p(\mu;V)$ (with Ω and m as above), then V can be embedded in each non-zero X_k in such a way that the element in X corresponding to an element $v\cdot\chi_B$ in $L^p(\mu;V)$ has the form $v\cdot\chi_D$ for some $D \subset \Omega$.

Proof: Let k be a point of Ω which is not an m-null point and D a clopen neighbourhood of k with finite m-measure. D corresponds to a projection E_B in $\mathfrak{A}$. For each v in V let f_v be the element of X corresponding to $v \cdot \chi_B$. Because m is the measure induced by μ, $\|f_v(.)\| = \|v\| \chi_D$, in particular $\|f_v(k)\| = \|v\|$. The mapping $v \mapsto f_v(k)$ is thus an isometric imbedding of V in X_k. It remains only to show that this mapping is independent of D. But if D' is another such clopen set, $E_{B'}$ the corresponding projection in $\mathfrak{A}$ and f'_v the element of X corresponding to $v \cdot \chi_{B'}$, then $E_B E_{B'}(v \cdot \chi_B - v\chi_{B'}) = 0$. It follows that $\chi_D \chi_{D'}(f_v - f'_v) = 0$. In particular, since $k \in D \cap D'$, $f_v(k) = f'_v(k)$. □

It might be expected that the imbedding of V in X_k described above is onto, but this is in general not the case. In fact one can show with the same methods employed in the next section in dealing with the general case that the imbedding is only onto if V is finite-dimensional or if k is an isolated point. In the case of finite-dimensional V we can identify X in 7.3 with the space $C^p_V(\Omega; m)$ of those continuous functions from Ω into the compactification $\overline{V}$ of V which are p-integrable with respect to m ($\overline{V}$ is obtained from the unit ball of V by identifying V with its interior by means of $v \leftrightarrow v/1+\|v\|$). The proof is an easy generalization of 3.4 .

As we mentioned earlier, it is of course possible that there are non-trivial L^p-projections in V itself. These then induce projections on $L^p(\mu; V)$ by virtue of $E \mapsto (f \mapsto E \circ f)$. We thus have two natural algebras of L^p-projections on $L^p(\mu; V)$, $\mathfrak{A}_1$ coming from the structure of μ and $\mathfrak{A}_2$ coming from the L^p-projections of V. It is now an obvious question as to whether one can construct the algebra of all L^p-projections out of $\mathfrak{A}_1$ and $\mathfrak{A}_2$ (in the case of p=2, a maximal algebra containing both $\mathfrak{A}_1$ and $\mathfrak{A}_2$). This question appears to be rather difficult to answer. Clearly some sort of product is called for. However we know that all three algebras , $\mathfrak{A}_1$, $\mathfrak{A}_2$, and the algebra of all L^p-projections, are complete. We therefore need a product which applied to two complete algebras gives a complete algebra again. Since it is known that the product of two extremally disconnected spaces is almost never extremally disconnected (see [S], p. 434) it is no use forming the Stonean spaces of $\mathfrak{A}_1$ and $\mathfrak{A}_2$. This leaves the measure-theoretical approach, i. e. to form the product of the measure spaces formed by the Stonean spaces of $\mathfrak{A}_1$ and $\mathfrak{A}_2$ together with a pair of perfect measures. This approach seems more promising but as far as we know there are no results in this direction as yet.

C: Some further properties of the component spaces

In the last section we have seen that for a finite-dimensional Banach space V all non-trivial components of a certain representation of $L^p(\mu ;V)$ as an integral module are isomorphic, $X_k \underset{\Phi}{\cong} X_l$, in such a way that for all $x_k \in X_k$ there is an $x \in X$ with $x(k) = x_k$ and $x(l) = \Phi x_k$. Thus the Banach space structure of X_k is transferred to X_l along suitable elements of the integral module, as it were. We also claimed that this does not hold for infinite-dimensional V. Now we shall examine how arbitrary integral modules behave in this respect.

<u>7.4 Definition</u>: Let $X \subset \int_K^p X_k \, dm$ be an integral module, $k \in K$, Y_k a subspace of X_k. A linear mapping $^-: Y_k \longrightarrow X$ is called an <u>almost isometric extension</u> of Y_k into X if

(1) $\overline{x_k}(k) = x_k$ for all $x_k \in Y_k$

(2) for all $\varepsilon > 0$ there is a neighbourhood C_ε of k with

$$(1-\varepsilon)\|x_k\| \le \|\overline{x_k}(\cdot)\|\big|_{C_\varepsilon} \le (1+\varepsilon)\|x_k\| \qquad \text{for all } x_k \in Y_k$$

<u>7.5 Proposition</u>: Let $X \subset \int_K^p X_k \, dm$ be an integral module, $k \in K$ not an m-null point. Then each finite-dimensional subspace of X_k has an almost isometric extension into X.

<u>Proof</u>: Let Y_k be an n-dimensional subspace of X_k, w.l.o.g. $n > 0$. Choose a clopen neighbourhood B of k with finite measure and $x^i \in X$ with $\|x^i(\cdot)\| = \chi_B$ and $(x^i(k))_{i=1}^n$ a basis of Y_k. Define a linear mapping $^-: Y_k \longrightarrow X$ by $x^i(k) \longmapsto x^i$. We set $S := \{s \mid s \in Y_k, \|s\| = 1\}$ and, for $\varepsilon > 0$, look for a neighbourhood C_ε of k with

$1 - \varepsilon \le \|\overline{s}(j)\| \le 1+\varepsilon$ for all $j \in C_\varepsilon$, $s \in S$.

As S is compact, it suffices to show the continuity of the mapping $(s,j) \longmapsto \|\overline{s}(j)\|$ at all points (s,k). For this purpose we may assume that $\|\sum_i \alpha_i x^i(k)\| = \sum_i |\alpha_i|$, as all norms induce the same topology on Y_k. If $s = \sum_i \alpha_i x^i(k) \in S$, $\varepsilon' > 0$, there is a clopen neighbourhood $D \subset B$ of k on which $\|\overline{s}(\cdot)\|$ varies less than $\frac{\varepsilon'}{2}$.

For $j \in D$ and $t = \sum_i \beta_i x^i(k) \in S$ with $\sum_i |\beta_i - \alpha_i| \le \frac{\varepsilon'}{2}$ we get

$$|\|\overline{t}(j)\| - \|\overline{s}(k)\|| \le |\|\overline{t}(j)\| - \|\overline{s}(j)\|| + |\|\overline{s}(j)\| - \|\overline{s}(k)\|| \le$$

$\le \|\overline{t-s}(j)\| + \frac{\varepsilon'}{2} \le \sum_i \|(\beta_i - \alpha_i)x^i(j)\| + \frac{\varepsilon'}{2} \le \varepsilon'$, which finishes the proof. □

Proposition 7.3 gives examples of infinite-dimensional subspaces of

X_k which can be almost isometrically extended into X. The next proposition shows that this does not hold for the space X_k itself (apart from trivial cases). As a corollary we get that the embedding of V into the component X_k (k non-isolated) of the representation of $L^p(\mu;V)$ given in the last section is not onto, since the subspace V of X_k is almost isometrically extendable.

<u>7.6 Proposition</u>: Let $X \subset \int_K^p X_k \, dm$ be an integral module, k a non-isolated point of K. If X_k is infinite-dimensional, it has no almost isometric extension into X.

<u>Proof</u>: We assume there is such an extension $^-$ and choose an arbitrary $\varepsilon \in]0,1[$. W.l.o.g. the associated clopen C has finite measure (k is not an m-null point). As $\{k\}$ is nowhere dense, the decreasing net (C_α) of clopen subsets of C containing k has $\emptyset$ as its infimum. If we choose an $x \in X$ with $\|x(\cdot)\| = \chi_C$, 1.6(ii) gives that the net $(m(C_\alpha)) = (\|\chi_{C_\alpha} x\|^{1/p})$ converges to zero. Therefore we can choose a decreasing sequence $(C_\nu)_{\nu \in \mathbb{N}}$ of clopen neighbourhoods C_ν of k contained in C with $m(C_\nu) \leq \nu^{-p}$ and set $R_\nu := C_\nu \setminus C_{\nu+1}$ $(\nu \in \mathbb{N})$.

We show that each neighbourhood B of k has a non-empty intersection with infinitely many R_ν's : -

For $\mu \in \mathbb{N}$ $B \cap C_\mu$ has positive measure. Choose $\rho > \mu$ with $\rho^{-p} < m(B \cap C_\mu)$; then $B \cap C_\mu$ is not contained in C_ρ. Hence $(B \cap C_\mu) \cap \bigcup_{\nu=\mu}^{\rho-1} R_\nu = (B \cap C_\mu) \setminus C_\rho \neq \emptyset$ and we have a $\nu \geq \mu$ with $B \cap R_\nu = (B \cap C_\mu) \cap R_\nu \neq \emptyset$.

Further we choose a sequence $(x_k^\nu)_{\nu \in \mathbb{N}}$ in the unit ball of X_k with $\|x_k^\nu - x_k^\mu\| > \frac{1}{2}$ for $\nu \neq \mu$. We define a sequence in X by $x^\nu := \sum_{\rho=1}^{\nu} \chi_{R_\rho} \overline{x_k^\rho}$, which is Cauchy because for $\mu > \nu$ we have $\|x^\mu - x^\nu\| =$

$(\int\limits_K \| \sum\limits_{\rho=\nu+1}^{\mu} \chi_{R_\rho} \overline{x_k^\rho}(\cdot)\|^p \, dm)^{1/p} \leq (1+\varepsilon)m(C_\nu)^{1/p} \leq \frac{1+\varepsilon}{\nu}$.

For $x := \lim\limits_\nu x^\nu$ evidently $x|_{R_\nu} = \overline{x_k^\nu}|_{R_\nu}$ $(\nu \in \mathbb{N})$. Further $x(k) \neq \infty$, since otherwise $\|x(\cdot)\||_B \geq 2$ on a suitable neighbourhood B of k and therefore $\|x(\cdot)\||_{B\cap R_\nu} = \|\overline{x_k^\nu}(\cdot)\||_{B\cap R_\nu} \geq 2$ for a suitable $\nu \in \mathbb{N}$ with $B\cap R_\nu \neq \emptyset$, which would contradict $\|\overline{x_k^\nu}(\cdot)\||_C \leq 1+\varepsilon$. □

We have $\infty \neq x(k) = \overline{x(k)}(k)$ and choose a neighbourhood D of k with $\|(x - \overline{x(k)})(\cdot)\||_D < \frac{1-\varepsilon}{4}$. Then for all $j \in R_\nu \cap D$ $\|x_k^\nu - x(k)\| \leq$ $\frac{1}{1-\varepsilon} \|(\overline{x_k^\nu} - \overline{x(k)})(j)\| = \frac{1}{1-\varepsilon}\|(x - \overline{x(k)})(j)\| \leq \frac{1}{1-\varepsilon} \cdot \frac{1-\varepsilon}{4} = \frac{1}{4}$, hence $\|x_k^\nu - x_k^\mu\| \leq \frac{1}{2}$ for suitable ν, μ $(\nu \neq \mu)$, which contradicts the choice of $(x_k^\nu)_{\nu \in \mathbb{N}}$.

The almost isometric extendability of finite-dimensional subspaces has an immediate corollary. We say a Banach space Z is <u>finitely representable</u> in a Banach space X or X <u>mimics</u> Z if for every finite-dimensional subspace Y of Z and every $\varepsilon > 0$ there is an injective linear operator T from Y into X with $|1 - \|T\| \cdot \|T^{-1}\|| \leq \varepsilon$.

Mimicry is a reflexive, transitive, but neither symmetric nor anti-symmetric relation (see, e.g. [D1] or [D2]).

<u>7.7 Corollary</u>: An integral module mimics each of its components.

<u>Proof</u>: Let Y_k be a finite-dimensional subspace of X_k, $\varepsilon > 0$. Choose the mapping $^-$ and C_ε as in 7.5 and set $Tx := m(C_\varepsilon)^{-1} \chi_{C_\varepsilon} \cdot \overline{x}$. □

The preceding results have a bearing on the following problem.

If we are given a Banach space X and a complete Boolean algebra $\mathfrak{A}$ of L^p-projections on it, we can represent X as an integral module over the Stonean space Ω of $\mathfrak{A}$ w.r.t. any perfect measure. If the point k is not an intrinsic null point (see 3.11) there is, by proposition 3.12, a representation in which X_k is not the null space and this space is independent of the representation. X and $\mathfrak{A}$ thus determine a mapping from $\Omega \setminus N$ (N the set of intrinsic null points) into a sufficiently large set of equivalence classes (under isometry) of non-trivial Banach spaces. It is natural to ask what properties such a mapping must have to arise in this way. If we could give necessary and sufficient conditions for this to be so, we should also solve the problem of the existence of an integral module in a given direct integral – $\int_K^p X_k \, dm$ would contain an integral module if and only if the mapping $k \longmapsto X_k$ defined on $K \setminus N_m$ (N_m the m-null points) were the restriction of a mapping defined on $K \setminus N$ satisfying these conditions. At present this problem is open. However, we obtain some necessary conditions on the dimensions of the components from proposition 7.5.

<u>7.8 Proposition</u>: Let X and $\mathfrak{A}$ be as above. The mapping $k \longmapsto \dim X_k$ from $\Omega \setminus N$ into $\mathbb{N} \cup \{\infty\}$ is continuous w.r.t. the relative and order topologies on $\Omega \setminus N$ and $\mathbb{N} \cup \{\infty\}$.

<u>Proof</u>: As $\Omega \setminus N$ is the union of the open sets $\Omega \setminus N_m$, we show the continuity of the above mapping defined on $\Omega \setminus N_m$ for each perfect measure m. For all $n \in \mathbb{N}$ the set $\{j \mid \dim X_j \geq n\}$ is open by 7.5. Hence it suffices to show that the sets $\{j \mid 0 < \dim X_j \leq n\}$ are open for all $n \in \mathbb{N}$.

Let $\dim X_k = n$ and choose $^-$ and C_ε for $Y_k := X_k$ and $\varepsilon := \frac{1}{4}$ as in 7.5. Clearly $\Phi_l(k) := \overline{x_k}(l)$ defines a one-to-one linear operator from X_k into X_l for all $l \in C_\varepsilon$. As k is not an element of the open set $M := \{l \mid l \in C_\varepsilon, \dim X_l > n\} = \{l \mid l \in C_\varepsilon, \Phi_l \text{ is not onto}\}$, it suffices to show that M is closed.

We apply the existence lemma to the clopen set $B := M^-$ and

$A := \{x \mid x \in X, \|x(\cdot)\| = \chi_D, D \subset B, d(x(l), \Phi_l(S)) \geq \frac{1}{2}$ for all $l \in D\}$

where S is defined as in 7.5. (ii) and (iii) are trivially satisfied.

Let C be a non-void subset of B. Then there is a j in $C \cap M$ and an $x \in X$ with $\|x(\cdot)\| = \chi_B$, $d(x(j), \Phi_j(S)) \geq \frac{3}{4}$. As in 7.5 by continuity and compactness we find a $D \subset C$ with $\|(x-\bar{s})(l)\| \geq \frac{1}{2}$ for all $l \in D$, $s \in S$. So we have (i), too. By the existence lemma there is an $x \in A$ with $\|x(\cdot)\| = \chi_B$. For all $l \in B$ we have

$\|x(l) - \frac{\bar{s}(l)}{\|\bar{s}(l)\|}\| \geq \frac{1}{2} - \|\bar{s}(l) - \frac{\bar{s}(l)}{\|\bar{s}(l)\|}\| \geq \frac{1}{4}$ and $\|x(l)\| = 1$, so that $x(l) \notin \Phi_l(X_k)$. $M = B$. □

As non-trivial components may accumulate at trivial ones, we had to exclude the nowhere dense set N from the domain of definition in order to get the continuity of dim. Thus we have lost the compactness of the domain of definition and we cannot conclude the closedness of dim. In fact, dim is in general neither open nor closed, as the example $X = \Pi^p_{n \in \mathbb{N}} X_n$ shows, where $X_n = \mathbb{R}^n$ for even n and $X_n = l^\infty$ for odd n (consider the integral module representation in 7.1 and the clopen subsets $2\mathbb{N}$ and $2\mathbb{N}-1$ of $\mathbb{N}$).

As an immediate consequence of the last proposition we get that the sets $\{k \mid \dim X_k = n\}$ are clopen in $K \setminus N$ for all $n \in \mathbb{N}$. Example 5.9 shows that this does not hold for $\{k \mid X_k$ separable$\}$ – the X_n, $n \in \mathbb{N}$, are finite-dimensional, but the components X_k, $k \in \beta\mathbb{N} \setminus \mathbb{N}$, contain l^∞ as a subspace and thus are not separable.

With the aid of 7.5 we get some further necessary conditions for the existence of integral modules in given direct integrals, which are of a more geometric nature, e.g. the continuity of the modulus of convexity (for all $\varepsilon \in [0,2]$ the mapping $k \longmapsto \delta_k(\varepsilon) :=$

$$:= \inf \{1 - \frac{\|x_k+y_k\|}{2} \mid x_k, y_k \in X_k, \|x_k\| = \|y_k\| = 1, \|x_k - y_k\| > \varepsilon\}$$

is continuous on $K \setminus N_m$).

At the end of chapter 5 we showed that for an integral module X the reflexivity of m-almost all components implies the reflexivity of X. However, we could not prove the converse statement. 7.5 and 7.7 give some partial answers under stronger assumptions, e.g. if X is uniformly convex as well, then all components X_k are also uniformly convex, hence reflexive. In this case we have also a representation of X' as a p'-integral module on K with components X_k' (see 5.8).

Appendix 1: The commutativity of L^p-projections

(outline of the proof; a detailed proof is given in [B2])

The essential idea in the proof of theorem 1.3 is the reduction to a special case:

Proposition: Let X be a Banach space, $1 \le p,q \le \infty$, $X = J_1 \oplus_p J_1^\perp = J_2 \oplus_q J_2^\perp$. Then $J := J_1 \cap J_2 + J_1 \cap J_2^\perp + J_1^\perp \cap J_2 + J_1^\perp \cap J_2^\perp$ is a closed subspace of X. If E_1 (resp. E_2) is the L^p-projection (resp. L^q-projection) onto J_1 (resp. J_2), then for i = 1,2 the mapping $\hat{E}_i : X/J \to X/J$, $\hat{E}_i([x]) := [E_i x]$ is well-defined and an L^p-projection (resp. L^q-projection). Let $\hat{J}_1 \oplus_p \hat{J}_1^\perp$ and $\hat{J}_2 \oplus_q \hat{J}_2^\perp$ be the decompositions of X/J defined by $\hat{E}_1$ and $\hat{E}_2$, respectively. We then have $\hat{J}_1 \cap \hat{J}_2 + \hat{J}_1 \cap \hat{J}_2^\perp + \hat{J}_1^\perp \cap \hat{J}_2 + \hat{J}_1^\perp \cap \hat{J}_2^\perp = 0$.

Definition: Let X be a Banach space, $1 \le p,q \le \infty$, $X = J_1 \oplus_p J_1^\perp = J_2 \oplus_q J_2^\perp$. We say that $J_1 \oplus_p J_1^\perp$, $J_2 \oplus_q J_2^\perp$ constitute a (p,q)-star, if $J_1 \cap J_2 + J_1 \cap J_2^\perp + J_1^\perp \cap J_2 + J_1^\perp \cap J_2^\perp = 0$.

Proposition: Let p,q, $1 \le p,q \le \infty$, be fixed such that there are no nontrivial (p,q)-stars (a (p,q)-star is called trivial if X = 0). Then for any Banach space X, any L^p-projection E_1 on X, and any L^q-projection E_2 on X we have $E_1E_2 = E_2E_1$. Also, if $p \ne q$, E_1 or E_2 must be trivial (that is 0 or Id). Conversely, as is easy to see, the commutativity of L^p-projections and L^q-projections implies that all (p,q)-stars are trivial. In particular, there are no nontrivial (1,1)-stars and (∞,∞)-stars.

Because of the preceding proposition theorem 1.3 will be proved if we are able to show that there are no nontrivial (p,q)-stars for $(p,q) \neq (2,2)$ (for $\{p,q\} \neq \{1,\infty\}$; if $\{p,q\} = \{1,\infty\}$ we have to prove thar $(\mathbb{R}^2, \| \ \|_1)$ is, up to isometric isomorphism, the only nontrivial space with $(1,\infty)$-stars).

To prove this we consider the following cases:

<u>$(p,q) = (1,1)$ or $(p,q) = (\infty,\infty)$</u>

We already noted that the commutativity of L^1-projections (resp. L^∞-projections) implies the nonexistence of nontrivial (1,1)-stars (resp. (∞,∞)-stars).

<u>$1 \leq p < \infty$, $q = \infty$</u>

Let $X = J_1 \oplus_p J_1^\perp = J_2 \oplus_\infty J_2^\perp$ be a (p,∞)-star, $X \neq 0$. We choose $x_0 \in J_1$, $\|x_0\| = 1$ and write $x_0 = ay + by^\perp$ whereby $a, b > 0$, $\max\{a,b\} = 1$ (w. l. o. g. $b = 1$), $y \in J_2$, $y^\perp \in J_2^\perp$, $\|y\| = \|y^\perp\| = 1$. Decomposing y, $y^\perp$ with respect to $X = J_1 \oplus_p J_1^\perp$ we get $y = \sqrt[p]{\lambda_1}\, x + \sqrt[p]{1-\lambda_1}\, x^\perp$, $y^\perp = \sqrt[p]{\lambda_2}\, \overline{x} - \sqrt[p]{1-\lambda_2}\, x^\perp$, whereby $0 < \lambda_1, \lambda_2 < 1$, $x, \overline{x} \in J_1$, $x^\perp \in J_1^\perp$, $\|x\| = \|\overline{x}\| = \|x^\perp\| = 1$. A detailed discussion of the three-dimensional subspace generated by x, $\overline{x}$, $x^\perp$ shows that a is necessarily equal to one (otherwise the unit ball of the linear hull of x, $\overline{x}$ would not be convex). So every $x_0 \in J_1$, $\|x_0\| = 1$, has a representation $x_0 = y + y^\perp$, whereby $y \in J_2$, $y^\perp \in J_2^\perp$, $\|y\| = \|y^\perp\| = 1$. We conclude a posteriori that $\lambda_1 = \lambda_2 = 1/2$ and $1 = (1/2)\|x - \overline{x}\|^p + 2^{p-1}$ (which already settles the case $p > 1$). As a consequence of the equality $x = \overline{x}$ we can restrict the decomposition $X = J_1 \oplus_1 J_1^\perp = J_2 \oplus_\infty J_2^\perp$ to finite dimensional subspaces. By showing that there are no $(1,\infty)$-stars on any four-dimensional Banach space we obtain a contradiction. To complete our argument it is necessary to prove that every nontrivial decomposition of X as $X = J_1 \oplus_1 J_1^{\perp} = J_2 \oplus_{\infty} J_2^{\perp}$ is already a star.

$1 < p \leqq \infty,\ q = 1$

By dualizing nontrivial (p,q)-stars and factorizing we obtain nontrivial (p',q')-stars. Hence, the foregoing discussion leads to the desired result for this case.

$1 < p,q < \infty$

In this case differentiability considerations of the norm on a fourdimensional subspace lead to the desired result. First, we proceed as in the case $1 \leqq p < \infty$, $q = \infty$, that is we start by decomposing an $x_0 \in J_1$ with respect to $J_2 \oplus_q J_2^{\perp}$. This time, however, we decompose the x, $\overline{x}$, $x^{\perp}$ which we get by this procedure once more with respect to $J_2 \oplus_q J_2^{\perp}$. This leads to a fourdimensional subspace in which we investigate the norm of $ax + b\overline{x} + cx^{\perp}$ for $a,b,c \in R$. We get $\|ax + b\overline{x}\|^p + |c|^p = F(a,b,c)$, where F is a function of a,b,c which is, as a function of c, infinetely often differentiable at $c = 0$ for certain fixed values of a and b. This fact implies that $|c|^p$ is infinitely often differentiable, hence $p \in 2\mathbb{N}$, in particular $p \geqq 2$. Because of the symmetry in p and q we also have $q \geqq 2$, and dualizing we get $p', q' \geqq 2$, too. Therefore $p = q = 2$ and the theorem is proved.

Appendix 2: L^∞-summands in CK-spaces

It is proved in [AE], prop. 6.18, that the L^∞-summands in CK (K a compact space) are precisely the annihilators of clopen subsets of K. This result is obtained as a corollary to a theorem which describes the M-ideals in C^*-algebras. Although we shall not make any major use of the result (cf. the counter-examples in section 1) we include a direct proof.

Lemma: If $E: CK \to CK$ is an L^∞-projection, then for $f,g \in CK$ $E(fg) = fE(g)$

Proof: Define $M_f: CK \to CK$ by $g \to fg$. We prove that the mappings M_f' and E' commute (which implies $M_f E = E M_f$). Because of 1.4 and the remark after the proof of 1.4 it is sufficient to show that M_f' can be arbitrarily well approximated by linear combinations of L^1-projections. For every Borel subset $A \subset K$ the operator $D_A: CK' \to CK'$, $m \to (g \to \int_A g dm)$, is in $\mathbb{P}_1$, and for $\| f - \sum_1^n a_i \chi_{A_i} \| \leq \epsilon$ we have $\| M_f' - \sum_1^n a_i D_{A_i} \| \leq \epsilon$ which is such an approximation.

Corollary: Every L^∞-summand J is a closed ideal in CK

Proof: Because of 1.2 an L^∞-summand is always closed. If $E^\perp$ is the L^∞-projection onto $J^\perp$, we have $J = \ker E^\perp$. It follows that J is an ideal.

Proposition: Let J be an L^∞-summand in CK. There is a clopen subset $C \subset K$ with $J = J_C := \{ f \mid f \in CK, f|_C = 0 \}$

Proof: The corollary implies the existence of closed subsets $C, C^\perp \subset K$ with $J = J_C$, $J^\perp = J_{C^\perp}$. We necessarily have $C \cup C^\perp = K$, $C \cap C^\perp = \emptyset$, because $J \oplus J^\perp = CK$. Thus C and $C^\perp$ are clopen.

<u>Corollary</u>: Let X be the space of bounded continuous functions on T, where T is a completely regular topological space. Then the L^∞-summands of X are in one-to-one correspondence with the clopen subsets of the Stone-Čech-compactification of T.

Appendix 3: A measure-theoretical approach to integral modules

In chapter 3 we saw how a Banach space could be represented as an integral of Banach spaces over a hyperstonean space (namely the Stonean space of a suitable Boolean algebra of L^p-projections). In this appendix we describe how the same can be done over an (almost) arbitrary measure space and show the relationship between the two constructions.

Suppose that X is a Banach space with a complete Boolean algebra $\mathfrak{A}$ of L^p-projections and that $L^p(S,\Sigma,\mu)$ is an L^p-space whose algebra of pseudo-characteristic projections is isomorphic to $\mathfrak{A}$. For each x in X we can define, in a similar way to the m_x of chapter 3, a measure μ_x on (S,Σ) by means of $\mu_x(C) := \| E_C x\|^p$ for C in Σ, whereby E_C is the projection in $\mathfrak{A}$ corresponding to $f \mapsto \chi_C f$ on $L^p(S,\Sigma,\mu)$. μ_x is a finite positive measure which is absolutely continuous with respect to μ, thus the Radon-Nikodym theorem applies and μ_x has a derivative in $L^1(S,\Sigma,\mu)$. The mapping $x \mapsto (d\mu_x/d\mu)^{1/p}$ is then a norm resolution for X taking values in $L^p(S,\Sigma,\mu)$. This norm resolution is related to that of chapter 3 in the following way. Let Ω and m be as in chapter 3. It follows from proposition 4.7, theorem 4.6 and the fact that $\mathfrak{A}$ and the algebra of pseudocharacteristic projections on $L^p(S,\Sigma,\mu)$ are isomorphic, that $L^p(S,\Sigma,\mu) = L^p(m)$. The norm resolution constructed above is then merely the composition of this isometry with the resolution of chapter 3.

If the measure space (S,Σ,μ) is strictly localizable then we can continue the construction to obtain an integral representation of X over (S, Σ, μ). Indeed if (S, Σ, μ) is strictly localizable then there is a lifting of $L^p(S, \Sigma, \mu)$ into $\mathcal{L}^p(S, \Sigma, \mu)$. By composing the norm resolution with the lifting we obtain a resolution whose values are functions instead of equivalence classes and can then proceed exactly as in chapter 3 to construct component spaces. Again the two methods are closely connected. A lifting of $L^p(S,\Sigma,\mu)$ defines a topology on S, the dense topology associated with the lifting. After identifying points in S not seperated by the topology, S thus topologized has a natural embedding as a dense subset of Ω. The representation of X obtained by restricting the integral module representation to the embedded S then turns out to be the same as that constructed with the help of the lifting.

Notation Index

$\mathfrak{A}$	an arbitrary Boolean algebra of L^p-projections, 48
$\mathbb{P}_p(X)$, $\mathbb{P}_p$	the set of all L^p-projections on X, 9
$E \wedge F$, $E \vee F$, $\overline{E}$	the inf, the sup, and the complementation in a Boolean algebra
$S(x;\mathfrak{A})$	the cycle generated by x, 47
lin	the linear hull of a subset
$\oplus$	the algebraic direct sum of two subspaces
$\oplus_p$	the abbreviation for direct p-decomposition, 5
X'	the space of continuous linear forms on the normed space X
$[X]$	the space of all continuous endomorphisms on the normed space X
$(\mathfrak{A})_{COMM}$	the commutator in $[X]$ of a subset $\mathfrak{A}$ of $[X]$
$(\mathfrak{A})^t$	the set of adjoints of the elements of $\mathfrak{A}$, $\mathfrak{A}$ a subset of $[X]$, 61
Id	the identity in $[X]$
$\cong$	isometric isomorphism
$P[X]$	the set of all projections in $[X]$ with norm ≤ 1, 13
$\lVert\cdot\rVert_p$	the L^p-norm
p'	the conjugate exponent to p, 8
$\prod\limits_{i\in I}^{p} X_i$	the p-direct product of a family of Banach spaces, 19

Ω, Ω_p	the Stonean space of a Boolean algebra, 3, 16, 48
βK	the Stone-Čech compactification of the completely regular space K
$C(K)$	the space of all continuous real-valued functions on K, K a topological space
χ_D	the characteristic function of the subset D
$C^p(K;m)$	the set of continuous p-integrable numerical valued functions on K, 32
$\dot{\cup}$	disjoint union
Δ	symmetric difference
$\setminus$	difference of two sets
$\mathbb{R}_+$	the set of nonnegative real numbers
[a,b]	the closed interval with the endpoints a and b
supp m	the support of the measure m
m_x	the measure on Ω which is generated by x, 24
$\int \lambda dE_\lambda$	the operator valued integral over the spectral family $(E_\lambda)_{\lambda\in\mathbb{R}}$
$d_{x,y}$, $\mu_{x,y}$	cf. lemma 5.3, 63
$J^\perp$	the L^p-summand orthogonal to J, 5
$C_p(X)$	the Cunningham p-algebra, 16
$C_p(X)^+$	the positive elements in $C_p(X)$, 18
$\leq_{C_p(X)^+}$ or $\leq_p$	the order in $C_p(X)$, 18
E_B	the L^p-projection associated with B, B clopen in Ω

B_E	the clopen set in Ω associated with the L^p-projection E
ψ	the isomorphism between $C_p(X)$ and $C(\Omega_p)$, 17
Φ	the integral module representation, 27
$\int_K^p X_k \, dm$	the p-direct integral of a family of Banach spaces, 33
$\langle x \rangle$	the image of an element x in the integral module, 37
$[x]$	the norm resolution of x, 36
X_k	the component in the representation associated with k, 37
$N(\cdot)$	the norm in the integral module, 33
$\mathcal{T}$	the set of all steplike functions, 45

Abbreviations:

w.l.o.g.	without loss of generality
w.r.t.	with respect to

Subject Index

almost isometric extension 88

Bochner space 85
Boolean algebra 3
Borel measure 2

characteristic projection 56
complementary F-summand 14
complementary L^p-summand 5
Cunningham p-algebra 16
cycle 47

direct integral 33
distinguished sequence of partitions 74
dual integral modules 62

essential p-direct integral 34

faithful bilinear form 62
F-projection 14
F-summand 14

hyperstonean 26

ideal 47
integral module 35
intrinsic null point 41

L^p-projection 5
L^p-summand 5

M-cycle 47
M-ideal 47
m-null point 41

norm resolution 30

p-direct integral 33
perfect measure 26
p-integral module 35
p-product 19
(p,q)-star 94
projection 5
pseudocharacteristic function (PCF) 58
pseudocharacteristic projection (PCP) 58

regular content 3

steplike function 45,66
Stonean space 4
spectral family 73

trivial L^p-structure 7

References

A: Papers concerning L^p-structure

[AE] E. M. Alfsen- E. G. Effros: Structure in real Banach spaces
Ann. of Math. 96 (1972)
Part I: 98 - 128
Part II: 129 - 173

[B1] E. Behrends: Über die L^p-Struktur in A(K)-Räumen
Math. Zeitschrift 139 (1974), 15 - 22

[B2] E. Behrends: L^p-Struktur in Banachräumen
Studia Math. 55 (1976), 71 - 85

[B3] E. Behrends: L^p-Struktur in Banachräumen II
erscheint in Studia Math. 62

[CS] H. B. Cohen - F. E. Sullivan: Projections onto cycles in smooth reflexive Banach spaces
Pac. Journal of Math. 34 (1970), 355-364

[C1] F. Cunningham jr.: L-structure in L-spaces
Trans. of the AMS 95 (1960), 274-299

[C2] F. Cunningham jr.: M-structure in M-spaces
Proc. of the Cambr. Phil. Soc. 63 (1967), 613-629

[CER] F.Cunningham-E.G.Effros-N.M.Roy: M-structure in dual Banach spaces
Isr. Journal of Math. 14 (1973), 304-308

[DGM] R.Danckwerts-S.Göbel-K.Meyfarth: Über die Cunningham-∞-Algebra und den Zentralisator reeller Banachräume
Math. Ann. 220 (1976), 163-169

[E1] R. Evans: Projektionen mit Normbedingungen in reellen Banachräumen
Dissertation, Freie Universität Berlin 1974

[E2] R. Evans: A characterization of M-summands
Proc. of the Cambr. Phil Soc. 76 (1974), 157-159

[F] H. Fakhoury: Existence d'une projection continue ...
J. math. pure et appl. 53 (1974), 1-16

[Gr] P. Greim: L-Struktur einiger Operatorräume
P. Greim: Zur Dualität zwischen Integralmoduln und Funktionenmoduln (submitted to the Math. Zeitschrift)

[HSW] R.Holmes-D.Scranton-J.Ward: Approximation from the space of compact operators and other M-ideals
Duke Math. J. 42 (1975), 259-269

[H] B. Hirsberg: M-ideals in complex function spaces and algebras
Isr. J. of Math. 12 (1972), 133-146

[R] N.M. Roy: A characterization of square Banach spaces
Isr. J. of Math 17 (1974), 142-148

[S1] F.E. Sullivan: Norm characterization of real L^p-spaces
Bull. of the AMS 74 (1968), 153-154

[S2] F.E. Sullivan: Structure of real L^p-spaces
J. Math. Anal. and Appl. 32 (1970), 621-629

B: Other papers

[Ba1] W.G. Bade: On Boolean algebras of projections, and algebras of operators
Trans. of the AMS 80 (1955), 345-360

[Ba2] W.G. Bade: A multiplicity theory for Boolean algebras
Trans. of the AMS 92 (1959), 508-530

[D1] M. M. Day: Normed linear spaces (3. ed.)
Springer Verlag, Berlin 1973

[D2] M. M. Day: Mimicry in normed spaces
Lect. Notes in Math. 490 (1975), 91-106

[Di] J. Diestel: Geometry in Banach spaces
Lect. Notes in Math . 485 (1975)

[G] A. Grothendieck: Une charactérisation vectorielle-métrique des espaces L^1
Canadian J. of Math. 7 (1955), 552-561

[H1] P.R. Halmos: Lectures on Boolean algebras
Springer Verlag, Berlin 1974

[H2] P.R. Halmos: Measure Theory
Springer Verlag, Berlin 1974

[HPh] E. Hille- R.S. Phillips : Functional analysis and semigroups
AMS, Coll. Publ. Vol 31, 1948

[J] R.C. James: Reflexivity and the supremum of linear functionals
Isr. J. of Math. 13 (1972), 289-300

[L] H.E. Lacey: The isometric theory of classical Banach spaces
Springer Verlag, Berlin 1974

[P] A. L. Peressini: Ordered topological vector spaces
Harper & Row, New York 1967

[Sch] H.H. Schaefer: Banach Lattices and positive Operators
Springer Verlag, Berlin 1974

[S] Z. Semadeni: Banach spaces of continuous functions I
Monografie Matematyczne 55, Warszawa 1971

Vol. 460: O. Loos, Jordan Pairs. XVI, 218 pages. 1975.

Vol. 461: Computational Mechanics. Proceedings 1974. Edited by J. T. Oden. VII, 328 pages. 1975.

Vol. 462: P. Gérardin, Construction de Séries Discrètes p-adiques. »Sur les séries discrètes non ramifiées des groupes réductifs déployés p-adiques«. III, 180 pages. 1975.

Vol. 463: H.-H. Kuo, Gaussian Measures in Banach Spaces. VI, 224 pages. 1975.

Vol. 464: C. Rockland, Hypoellipticity and Eigenvalue Asymptotics. III, 171 pages. 1975.

Vol. 465: Séminaire de Probabilités IX. Proceedings 1973/74. Edité par P. A. Meyer. IV, 589 pages. 1975.

Vol. 466: Non-Commutative Harmonic Analysis. Proceedings 1974. Edited by J. Carmona, J. Dixmier and M. Vergne. VI, 231 pages. 1975.

Vol. 467: M. R. Essén, The Cos $\pi\lambda$ Theorem. With a paper by Christer Borell. VII, 112 pages. 1975.

Vol. 468: Dynamical Systems – Warwick 1974. Proceedings 1973/74. Edited by A. Manning. X, 405 pages. 1975.

Vol. 469: E. Binz, Continuous Convergence on C(X). IX, 140 pages. 1975.

Vol. 470: R. Bowen, Equilibrium States and the Ergodic Theory of Anosov Diffeomorphisms. III, 108 pages. 1975.

Vol. 471: R. S. Hamilton, Harmonic Maps of Manifolds with Boundary. III, 168 pages. 1975.

Vol. 472: Probability-Winter School. Proceedings 1975. Edited by Z. Ciesielski, K. Urbanik, and W. A. Woyczyński. VI, 283 pages. 1975.

Vol. 473: D. Burghelea, R. Lashof, and M. Rothenberg, Groups of Automorphisms of Manifolds. (with an appendix by E. Pedersen) VII, 156 pages. 1975.

Vol. 474: Séminaire Pierre Lelong (Analyse) Année 1973/74. Edité par P. Lelong. VI, 182 pages. 1975.

Vol. 475: Répartition Modulo 1. Actes du Colloque de Marseille-Luminy, 4 au 7 Juin 1974. Edité par G. Rauzy. V, 258 pages. 1975. 1975.

Vol. 476: Modular Functions of One Variable IV. Proceedings 1972. Edited by B. J. Birch and W. Kuyk. V, 151 pages. 1975.

Vol. 477: Optimization and Optimal Control. Proceedings 1974. Edited by R. Bulirsch, W. Oettli, and J. Stoer. VII, 294 pages. 1975.

Vol. 478: G. Schober, Univalent Functions – Selected Topics. V, 200 pages. 1975.

Vol. 479: S. D. Fisher and J. W. Jerome, Minimum Norm Extremals in Function Spaces. With Applications to Classical and Modern Analysis. VIII, 209 pages. 1975.

Vol. 480: X. M. Fernique, J. P. Conze et J. Gani, Ecole d'Eté de Probabilités de Saint-Flour IV-1974. Edité par P.-L. Hennequin. XI, 293 pages. 1975.

Vol. 481: M. de Guzmán, Differentiation of Integrals in R^n. XII, 226 pages. 1975.

Vol. 482: Fonctions de Plusieurs Variables Complexes II. Séminaire François Norguet 1974–1975. IX, 367 pages. 1975.

Vol. 483: R. D. M. Accola, Riemann Surfaces, Theta Functions, and Abelian Automorphisms Groups. III, 105 pages. 1975.

Vol. 484: Differential Topology and Geometry. Proceedings 1974. Edited by G. P. Joubert, R. P. Moussu, and R. H. Roussarie. IX, 287 pages. 1975.

Vol. 485: J. Diestel, Geometry of Banach Spaces – Selected Topics. XI, 282 pages. 1975.

Vol. 486: S. Stratila and D. Voiculescu, Representations of AF-Algebras and of the Group U (∞). IX, 169 pages. 1975.

Vol. 487: H. M. Reimann und T. Rychener, Funktionen beschränkter mittlerer Oszillation. VI, 141 Seiten. 1975.

Vol. 488: Representations of Algebras, Ottawa 1974. Proceedings 1974. Edited by V. Dlab and P. Gabriel. XII, 378 pages. 1975.

Vol. 489: J. Bair and R. Fourneau, Etude Géométrique des Espaces Vectoriels. Une Introduction. VII, 185 pages. 1975.

Vol. 490: The Geometry of Metric and Linear Spaces. Proceedings 1974. Edited by L. M. Kelly. X, 244 pages. 1975.

Vol. 491: K. A. Broughan, Invariants for Real-Generated Uniform Topological and Algebraic Categories. X, 197 pages. 1975.

Vol. 492: Infinitary Logic: In Memoriam Carol Karp. Edited by D. W. Kueker. VI, 206 pages. 1975.

Vol. 493: F. W. Kamber and P. Tondeur, Foliated Bundles and Characteristic Classes. XIII, 208 pages. 1975.

Vol. 494: A Cornea and G. Licea. Order and Potential Resolvent Families of Kernels. IV, 154 pages. 1975.

Vol. 495: A. Kerber, Representations of Permutation Groups II. V, 175 pages. 1975.

Vol. 496: L. H. Hodgkin and V. P. Snaith, Topics in K-Theory. Two Independent Contributions. III, 294 pages. 1975.

Vol. 497: Analyse Harmonique sur les Groupes de Lie. Proceedings 1973–75. Edité par P. Eymard et al. VI, 710 pages. 1975.

Vol. 498: Model Theory and Algebra. A Memorial Tribute to Abraham Robinson. Edited by D. H. Saracino and V. B. Weispfenning. X, 463 pages. 1975.

Vol. 499: Logic Conference, Kiel 1974. Proceedings. Edited by G. H. Müller, A. Oberschelp, and K. Potthoff. V, 651 pages 1975.

Vol. 500: Proof Theory Symposion, Kiel 1974. Proceedings. Edited by J. Diller and G. H. Müller. VIII, 383 pages. 1975.

Vol. 501: Spline Functions, Karlsruhe 1975. Proceedings. Edited by K. Böhmer, G. Meinardus, and W. Schempp. VI, 421 pages. 1976.

Vol. 502: János Galambos, Representations of Real Numbers by Infinite Series. VI, 146 pages. 1976.

Vol. 503: Applications of Methods of Functional Analysis to Problems in Mechanics. Proceedings 1975. Edited by P. Germain and B. Nayroles. XIX, 531 pages. 1976.

Vol. 504: S. Lang and H. F. Trotter, Frobenius Distributions in GL_2-Extensions. III, 274 pages. 1976.

Vol. 505: Advances in Complex Function Theory. Proceedings 1973/74. Edited by W. E. Kirwan and L. Zalcman. VIII, 203 pages. 1976.

Vol. 506: Numerical Analysis, Dundee 1975. Proceedings. Edited by G. A. Watson. X, 201 pages. 1976.

Vol. 507: M. C. Reed, Abstract Non-Linear Wave Equations. VI, 128 pages. 1976.

Vol. 508: E. Seneta, Regularly Varying Functions. V, 112 pages. 1976.

Vol. 509: D. E. Blair, Contact Manifolds in Riemannian Geometry. VI, 146 pages. 1976.

Vol. 510: V. Poènaru, Singularités C^∞ en Présence de Symétrie. V, 174 pages. 1976.

Vol. 511: Séminaire de Probabilités X. Proceedings 1974/75. Edité par P. A. Meyer. VI, 593 pages. 1976.

Vol. 512: Spaces of Analytic Functions, Kristiansand, Norway 1975. Proceedings. Edited by O. B. Bekken, B. K. Øksendal, and A. Stray. VIII, 204 pages. 1976.

Vol. 513: R. B. Warfield, Jr. Nilpotent Groups. VIII, 115 pages. 1976.

Vol. 514: Séminaire Bourbaki vol. 1974/75. Exposés 453 – 470. IV, 276 pages. 1976.

Vol. 515: Bäcklund Transformations. Nashville, Tennessee 1974. Proceedings. Edited by R. M. Miura. VIII, 295 pages. 1976.

Vol. 516: M. L. Silverstein, Boundary Theory for Symmetric Markov Processes. XVI, 314 pages. 1976.

Vol. 517: S. Glasner, Proximal Flows. VIII, 153 pages. 1976.

Vol. 518: Séminaire de Théorie du Potentiel, Proceedings Paris 1972–1974. Edité par F. Hirsch et G. Mokobodzki. VI, 275 pages. 1976.

Vol. 519: J. Schmets, Espaces de Fonctions Continues. XII, 150 pages. 1976.

Vol. 520: R. H. Farrell, Techniques of Multivariate Calculation. X, 337 pages. 1976.

Vol. 521: G. Cherlin, Model Theoretic Algebra – Selected Topics. IV, 234 pages. 1976.

Vol. 522: C. O. Bloom and N. D. Kazarinoff, Short Wave Radiation Problems in Inhomogeneous Media: Asymptotic Solutions. V. 104 pages. 1976.

Vol. 523: S. A. Albeverio and R. J. Høegh-Krohn, Mathematical Theory of Feynman Path Integrals. IV, 139 pages. 1976.

Vol. 524: Séminaire Pierre Lelong (Analyse) Année 1974/75. Edité par P. Lelong. V, 222 pages. 1976.

Vol. 525: Structural Stability, the Theory of Catastrophes, and Applications in the Sciences. Proceedings 1975. Edited by P. Hilton. VI, 408 pages. 1976.

Vol. 526: Probability in Banach Spaces. Proceedings 1975. Edited by A. Beck. VI, 290 pages. 1976.

Vol. 527: M. Denker, Ch. Grillenberger, and K. Sigmund, Ergodic Theory on Compact Spaces. IV, 360 pages. 1976.

Vol. 528: J. E. Humphreys, Ordinary and Modular Representations of Chevalley Groups. III, 127 pages. 1976.

Vol. 529: J. Grandell, Doubly Stochastic Poisson Processes. X, 234 pages. 1976.

Vol. 530: S. S. Gelbart, Weil's Representation and the Spectrum of the Metaplectic Group. VII, 140 pages. 1976.

Vol. 531: Y.-C. Wong, The Topology of Uniform Convergence on Order-Bounded Sets. VI, 163 pages. 1976.

Vol. 532: Théorie Ergodique. Proceedings 1973/1974. Edité par J.-P. Conze and M. S. Keane. VIII, 227 pages. 1976.

Vol. 533: F. R. Cohen, T. J. Lada, and J. P. May, The Homology of Iterated Loop Spaces. IX, 490 pages. 1976.

Vol. 534: C. Preston, Random Fields. V, 200 pages. 1976.

Vol. 535: Singularités d'Applications Differentiables. Plans-sur-Bex. 1975. Edité par O. Burlet et F. Ronga. V, 253 pages. 1976.

Vol. 536: W. M. Schmidt, Equations over Finite Fields. An Elementary Approach. IX, 267 pages. 1976.

Vol. 537: Set Theory and Hierarchy Theory. Bierutowice, Poland 1975. A Memorial Tribute to Andrzej Mostowski. Edited by W. Marek, M. Srebrny and A. Zarach. XIII, 345 pages. 1976.

Vol. 538: G. Fischer, Complex Analytic Geometry. VII, 201 pages. 1976.

Vol. 539: A. Badrikian, J. F. C. Kingman et J. Kuelbs, Ecole d'Eté de Probabilités de Saint Flour V-1975. Edité par P.-L. Hennequin. IX, 314 pages. 1976.

Vol. 540: Categorical Topology, Proceedings 1975. Edited by E. Binz and H. Herrlich. XV, 719 pages. 1976.

Vol. 541: Measure Theory, Oberwolfach 1975. Proceedings. Edited by A. Bellow and D. Kölzow. XIV, 430 pages. 1976.

Vol. 542: D. A. Edwards and H. M. Hastings, Čech and Steenrod Homotopy Theories with Applications to Geometric Topology. VII, 296 pages. 1976.

Vol. 543: Nonlinear Operators and the Calculus of Variations, Bruxelles 1975. Edited by J. P. Gossez, E. J. Lami Dozo, J. Mawhin, and L. Waelbroeck, VII, 237 pages. 1976.

Vol. 544: Robert P. Langlands, On the Functional Equations Satisfied by Eisenstein Series. VII, 337 pages. 1976.

Vol. 545: Noncommutative Ring Theory. Kent State 1975. Edited by J. H. Cozzens and F. L. Sandomierski. V, 212 pages. 1976.

Vol. 546: K. Mahler, Lectures on Transcendental Numbers. Edited and Completed by B. Diviš and W. J. Le Veque. XXI, 254 pages. 1976.

Vol. 547: A. Mukherjea and N. A. Tserpes, Measures on Topological Semigroups: Convolution Products and Random Walks. V, 197 pages. 1976.

Vol. 548: D. A. Hejhal, The Selberg Trace Formula for PSL (2,ℝ). Volume I. VI, 516 pages. 1976.

Vol. 549: Brauer Groups, Evanston 1975. Proceedings. Edited by D. Zelinsky. V, 187 pages. 1976.

Vol. 550: Proceedings of the Third Japan – USSR Symposium on Probability Theory. Edited by G. Maruyama and J. V. Prokhorov. VI, 722 pages. 1976.

Vol. 551: Algebraic K-Theory, Evanston 1976. Proceedings. Edited by M. R. Stein. XI, 409 pages. 1976.

Vol. 552: C. G. Gibson, K. Wirthmüller, A. A. du Plessis and E. J. N. Looijenga. Topological Stability of Smooth Mappings. V, 155 pages. 1976.

Vol. 553: M. Petrich, Categories of Algebraic Systems. Vector and Projective Spaces, Semigroups, Rings and Lattices. VIII, 217 pages. 1976.

Vol. 554: J. D. H. Smith, Mal'cev Varieties. VIII, 158 pages. 1976.

Vol. 555: M. Ishida, The Genus Fields of Algebraic Number Fields. VII, 116 pages. 1976.

Vol. 556: Approximation Theory. Bonn 1976. Proceedings. Edited by R. Schaback and K. Scherer. VII, 466 pages. 1976.

Vol. 557: W. Iberkleid and T. Petrie, Smooth S^1 Manifolds. III, 163 pages. 1976.

Vol. 558: B. Weisfeiler, On Construction and Identification of Graphs. XIV, 237 pages. 1976.

Vol. 559: J.-P. Caubet, Le Mouvement Brownien Relativiste. IX, 212 pages. 1976.

Vol. 560: Combinatorial Mathematics, IV, Proceedings 1975. Edited by L. R. A. Casse and W. D. Wallis. VII, 249 pages. 1976.

Vol. 561: Function Theoretic Methods for Partial Differential Equations. Darmstadt 1976. Proceedings. Edited by V. E. Meister, N. Weck and W. L. Wendland. XVIII, 520 pages. 1976.

Vol. 562: R. W. Goodman, Nilpotent Lie Groups: Structure and Applications to Analysis. X, 210 pages. 1976.

Vol. 563: Séminaire de Théorie du Potentiel. Paris, No. 2. Proceedings 1975–1976. Edited by F. Hirsch and G. Mokobodzki. VI, 292 pages. 1976.

Vol. 564: Ordinary and Partial Differential Equations, Dundee 1976. Proceedings. Edited by W. N. Everitt and B. D. Sleeman. XVIII, 551 pages. 1976.

Vol. 565: Turbulence and Navier Stokes Equations. Proceedings 1975. Edited by R. Temam. IX, 194 pages. 1976.

Vol. 566: Empirical Distributions and Processes. Oberwolfach 1976. Proceedings. Edited by P. Gaenssler and P. Révész. VII, 146 pages. 1976.

Vol. 567: Séminaire Bourbaki vol. 1975/76. Exposés 471–488. IV, 303 pages. 1977.

Vol. 568: R. E. Gaines and J. L. Mawhin, Coincidence Degree, and Nonlinear Differential Equations. V, 262 pages. 1977.

Vol. 569: Cohomologie Etale SGA 4½. Séminaire de Géométrie Algébrique du Bois-Marie. Edité par P. Deligne. V, 312 pages. 1977.

Vol. 570: Differential Geometrical Methods in Mathematical Physics, Bonn 1975. Proceedings. Edited by K. Bleuler and A. Reetz. VIII, 576 pages. 1977.

Vol. 571: Constructive Theory of Functions of Several Variables, Oberwolfach 1976. Proceedings. Edited by W. Schempp and K. Zeller. VI, 290 pages. 1977

Vol. 572: Sparse Matrix Techniques, Copenhagen 1976. Edited by V. A. Barker. V, 184 pages. 1977.

Vol. 573: Group Theory, Canberra 1975. Proceedings. Edited by R. A. Bryce, J. Cossey and M. F. Newman. VII, 146 pages. 1977.

Vol. 574: J. Moldestad, Computations in Higher Types. IV, 203 pages. 1977.

Vol. 575: K-Theory and Operator Algebras, Athens, Georgia 1975. Edited by B. B. Morrel and I. M. Singer. VI, 191 pages. 1977.

Vol. 576: V. S. Varadarajan, Harmonic Analysis on Real Reductive Groups. VI, 521 pages. 1977.

Vol. 577: J. P. May, E_∞ Ring Spaces and E_∞ Ring Spectra. IV, 268 pages. 1977.

Vol. 578: Séminaire Pierre Lelong (Analyse) Année 1975/76. Edité par P. Lelong. VI, 327 pages. 1977.

Vol. 579: Combinatoire et Représentation du Groupe Symétrique, Strasbourg 1976. Proceedings 1976. Edité par D. Foata. IV, 339 pages. 1977.